Introduction to Organic Chemistry II

11th Hour

Introduction to Organic Chemistry II

Seth Elsheimer

Department of Chemistry
University of Central Florida
Orlando, Florida

b

**Blackwell
Science**

Editorial Offices:
Commerce Place, 350 Main Street, Malden, Massachusetts 02148, USA
Osney Mead, Oxford OX2 0EL, England
25 John Street, London WC1N 2BL, England
23 Ainslie Place, Edinburgh EH3 6AJ, Scotland
54 University Street, Carlton, Victoria 3053, Australia
Other Editorial Offices:
Blackwell Wissenschafts-Verlag GmbH, Kurfürstendamm 57, 10707 Berlin, Germany
Blackwell Science KK, MG Kodenmacho Building, 7-10 Kodenmacho Nihombashi, Chuo-ku, Tokyo 104, Japan

Distributors:
USA
> Blackwell Science, Inc.
> Commerce Place
> 350 Main Street
> Malden, Massachusetts 02148
> (Telephone orders: 800-215-1000 or 781-388-8250; fax orders: 781-388-8270)

Canada
> Login Brothers Book Company
> 324 Saulteaux Crescent
> Winnipeg, Manitoba, R3J 3T2
> (Telephone orders: 204-837-2987)

Australia
> Blackwell Science Pty, Ltd.
> 54 University Street
> Carlton, Victoria 3053
> (Telephone orders: 03-9347-0300; fax orders: 03-9349-3016)

Outside North America and Australia
> Blackwell Science, Ltd.
> c/o Marston Book Services, Ltd.
> P.O. Box 269
> Abingdon
> Oxon OX14 4YN
> England
> (Telephone orders: 44-01235-465500; fax orders: 44-01235-465555)

Acquisitions: Nancy Hill-Whilton
Development: Jill Connor
Production: Louis C. Bruno, Jr.
Manufacturing: Lisa Flanagan
Interior design by Colour Mark
Cover design by Madison Design
Typeset by Best-set Typesetter Ltd., Hong Kong

00 01 02 03 5 4 3 2 1

The Blackwell Science logo is a trade mark of Blackwell Science Ltd., registered at the United Kingdom Trade Marks Registry

Library of Congress Cataloging-in-Publication Data

Elsheimer, Seth Robert.
    Introduction to organic chemistry II / by Seth Elshimer.
      p. cm.—(11th hour)
    ISBN 0-86542-317-2
      1. Chemistry, Organic—Outlines, syllabi, etc.   I. Title.   II. 11th hour (Malden, Mass.)
    QD256.5 .E39 2000
    547—dc21   00-022800

CONTENTS

11TH HOUR GUIDE TO SUCCESS

The 11th Hour Series is designed to be used when the textbook doesn't make sense, the course content is tough, or when you just want a better grade in the course. It can be used from the beginning to the end of the course for best results or when cramming for exams. Both professors teaching the course and students who have taken it have reviewed this material to make sure it does what *you* need it to do. The material flows so that the process keeps your mind actively learning. The idea is to cut through the fluff, get to what you need to know, and then help you understand it.

Essential Background. We tell you what information you already need to know to comprehend the topic. You can then review or apply the appropriate concepts to conquer the new material.

Key Points. We highlight the key points of each topic, phrasing them as questions to engage active learning. A brief explanation of the topic follows the points.

Topic Tests. We immediately follow each topic with a brief test so that the topic is reinforced. This helps you prepare for the real thing.

Answers. Answers come right after the tests; but, we take it a step farther (that reinforcement thing again), we explain the answers.

Clinical Correlation or Application. It helps immeasurably to understand academic topics when they are presented in a clinical situation or an everyday, real-world example. We provide one in every chapter.

Demonstration Problem. Some science topics involve a lot of problem solving. Where it's helpful, we demonstrate a typical problem with step-by-step explanation.

Chapter Test. For more reinforcement, there is a test at the end of every chapter that covers all of the topics. The questions are essay, multiple choice, short answer, and true/false to give you plenty of practice and a chance to reinforce the material the way you find easiest. Answers are provided after the test.

Check Your Performance. After the chapter test we provide a performance check to help you spot your weak areas. You will then know if there is something you should look at once more.

Sample Midterms and Final Exams. Practice makes perfect so we give you plenty of opportunity to practice acing those tests.

The Web. Whenever you see this symbol ▣ the author has put something on the Web page that relates to that content. It could be a caution or a hint, an illustration or simply more explanation. You can access the appropriate page through *http://www.blackwellscience.com*. Then click on the title of this book.

The whole flow of this review guide is designed to keep you actively engaged in understanding the material. You'll get what you need fast, and you will reinforce it painlessly. Unfortunately, we can't take the exams for you!

PREFACE

This book was designed and written with you in mind. The goal is to quickly review the most important parts of a standard second-semester organic chemistry course. The format and coverage are intentionally brief. I assume you have completed the first semester and have completed, or are currently enrolled in, a second-semester organic chemistry course. This book contains relatively little of the usual pedagogical material and detailed explanations found in a standard organic chemistry textbook. Only the most gifted student is likely to master the concepts using this book as the only information source. Where possible, the theoretical bases for observable phenomena are mentioned; however, the brief review format unavoidably requires less explaining and more summarizing. Those conscientious students seeking a more detailed analysis or theoretical background are encouraged to consult one of the many excellent organic chemistry textbooks available. There are also web supplements referenced throughout this book. Many of these contain more detailed explanations and more examples. These can be accessed via http://www.blackwellscience.com.

Chapters 1 and 2 (Aromatic Compounds, and Spectroscopy) are Chapters 8 and 9 from the *Introduction to Organic Chemistry I* book. This duplication is to accommodate variation in course content among different institutions.

Sincere thanks go to Nancy Hill-Whilton, Executive Editor, and Jill Connor, Assistant Development Editor for their frequent enthusiastic encouragement and guidance. Thanks also to Lou Bruno and the rest of the talented production staff at Blackwell Science, Inc. I gratefully acknowledge the following professors and students who reviewed the manuscript before publication and made helpful suggestions: Paul Buonora, University of Scranton; Patricia Heiden, Michigan Tech; Christopher Singh, Boston University; and Christy Kenny, University of Pennsylvania School of Medicine.

This book is dedicated to Janice Elsheimer—wife, author, and educator.

Seth Elsheimer, Ph.D.
Associate Professor
Department of Chemistry
University of Central Florida

CHAPTER 1

Aromatic Compounds

(This chapter is Chapter 8 in Elsheimer, S. [2000] *Introduction to Organic Chemistry I.* Blackwell Science, Malden, Massachusetts.)

At one time organic compounds were classified as "aromatic" based largely on their pleasant fragrances. As organic chemistry evolved, this term began to include compounds under this classification based on their structures and patterns of reactivity rather than their ability to stimulate an olfactory response. In this chapter we examine the structural criteria for aromaticity and look at some of the most important transformations of aromatic compounds.

ESSENTIAL BACKGROUND

- **Resonance**
- **Conventions for writing mechanisms**
- **Electrophilic addition**
- **Nucleophiles and leaving groups**

TOPIC 1: AROMATICITY, BENZENE, AND RESONANCE

KEY POINTS

✓ *What is meant by the term "aromatic"?*

✓ *What structural requirements define aromaticity?*

✓ *What is a Hückel number?*

✓ *What is the structure of benzene?*

✓ *What are some characteristic properties of aromatic compounds?*

To a modern organic chemist, the term **aromatic** has nothing to do with how a compound smells. Some aromatic compounds have no aroma at all. This term is used to describe compounds that are cyclic, planar, and completely conjugated around the ring and have a **Hückel number** of pi electrons. A Hückel number is given by $4n + 2$, where "n" is an integer ($n = 0, 1, 2, 3, . . .$), and therefore the Hückel numbers are 2, 6, 10, 14, ... A typical aromatic compound is benzene. Although a traditional Lewis-Kekule' representation of benzene would suggest it has alternating double and single bonds, experimentally it has been shown that all C—C bonds of benzene are the same length and strength. The apparent three pi bonds are not localized but rather are spread out over the entire ring. The C—C bonds are neither single nor double but rather are something between these extremes. This situation is shown below with resonance forms. Benzene and its derivatives are also represented with a circle signifying the six pi electrons of the ring.

Benzene and other aromatic compounds are unusually stable. The apparent pi bonds of benzene do not, in general, undergo addition reactions like those of alkenes but rather tend to undergo substitution reactions.

No reaction $\xleftarrow{\text{HBr}}$ $\xrightarrow[\text{FeBr}_3]{\text{Br}_2}$ (+ HBr)

Topic Test 1: Aromaticity, Benzene, and Resonance

True/False

1. 1,3-Cyclobutadiene is aromatic.

2. 1,3,5-Cycloheptatriene is aromatic.

Multiple Choice

3. Which of the following ions is aromatic?

d. All of the above
e. None of the above

4. Which of the following heterocyclic compounds is aromatic?

d. All of the above
e. None of the above

Short Answer

5. List the structural requirements for a compound to be aromatic.

6. 7-Iodo-1,3,5-cycloheptatriene can lose an iodide ion to form a relatively stable carbocation with the formula $C_7H_7^+$. Show the structure of this cation and explain its stability.

Topic Test 1: Answers

1. **False.** Although 1,3-cyclobutadiene is cyclic and completely conjugated and can be represented by two equivalent resonance forms, it has four pi electrons and therefore does not conform to the 4n + 2 rule and is not aromatic.

2. **False.** Although 1,3,5-cycloheptatriene is cyclic and has six pi electrons, it is not completely conjugated all the way around the ring. Carbon 7 is a saturated methylene, CH_2.

3. **b.** The cyclopentadienyl anion is aromatic. It is cyclic, completely conjugated around the ring, and has six pi electrons. Answers a and c each have four pi electrons.

4. **d.** In each of cases a, b, and c, the cyclic conjugated system contains six pi electrons. Knowing where the "nonbonded" or "lone pair" electrons reside is essential here.

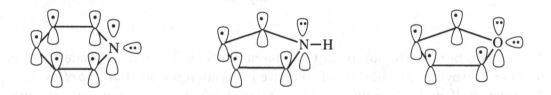

5. Aromatic molecules or ions must be cyclic, planar, and completely conjugated around the ring and must have a Hückel number (4n + 2) of pi electrons (2, 6, 10, 14, . . .).

6. The stable ion is a cycloheptatrienyl cation, which is aromatic.

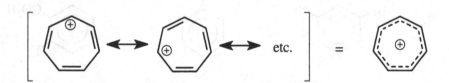

TOPIC 2: AROMATIC NOMENCLATURE

KEY POINTS

✓ *How are benzene derivatives named?*

✓ *What parent aromatic names are used to derive other names?*

✓ *What do the terms* ortho, meta, *and* para *mean with regard to substituents?*

✓ *How are benzene and related compounds numbered?*

Monosubstituted benzenes are named as such in many cases; however, there are some widely recognized **parent names** in which the single substituent on benzene gives the compound a different parent name.

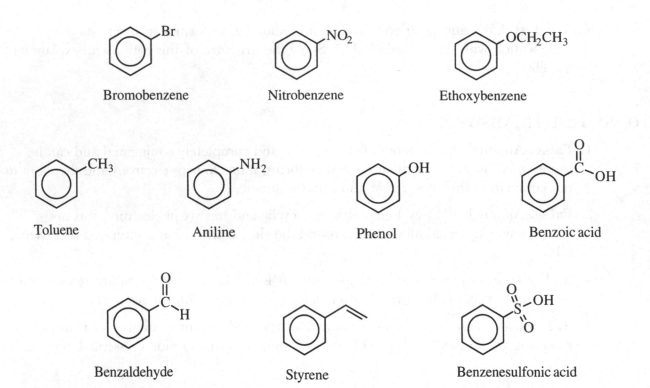

Bromobenzene   Nitrobenzene   Ethoxybenzene

Toluene   Aniline   Phenol   Benzoic acid

Benzaldehyde   Styrene   Benzenesulfonic acid

The carbon bearing the substituent in a monosubstituted benzene is understood to be carbon number 1. Any second substituent will have its location specified as 2 (**ortho**, *o*), 3 (**meta**, *m*), or 4 (**para**, *p*). If more than two substituents are attached to a benzene ring, the prefixes *o*, *m*, and *p* are not used and number locators are used exclusively. Substituents are listed in alphabetical order and the numbering is in the direction that gives the lowest numbers.

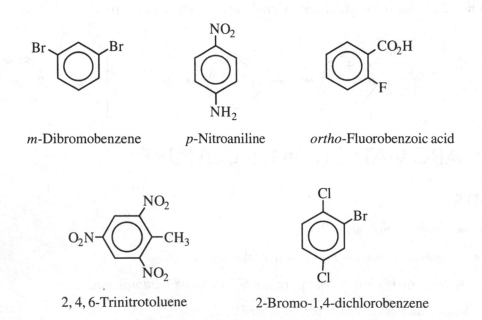

m-Dibromobenzene   *p*-Nitroaniline   *ortho*-Fluorobenzoic acid

2, 4, 6-Trinitrotoluene   2-Bromo-1,4-dichlorobenzene

The aromatic substituent group C_6H_5 is called **phenyl** (not benzyl) and is often abbreviated Ph. A **benzyl** group is a phenyl attached to a methylene (i.e., —CH_2Ph).

1,2,3,4-Tetraphenyl-1,3-cyclopentadiene

Benzyl iodide

Topic Test 2: Aromatic Nomenclature

Short Answer

1–3. Name the following.

1. [structure of benzene with I and CH₃ substituents]

2. [structure of phenyl-cycloheptadiene]

3. [structure of benzene with two Cl and OH]

Provide unambiguous structural formulas for the compounds named below.

4. *p*-Ethylbenzaldehyde

5. 2,4-Dibromostyrene

6. *m*-Dinitrobenzene

Topic Test 2: Answers

1. *ortho*-Iodotoluene (or *o*-iodotoluene or 2-iodotoluene)

2. 2-Phenyl-1,3-cycloheptadiene

3. 3,5-Dichlorophenol

4. CH_3CH_2—[benzene ring]—$\overset{O}{\underset{H}{C}}$

5. [structure of dibromostyrene with two Br]

6. [structure of benzene with two NO_2]

TOPIC 3: GENERAL EAS REACTION AND SOME SPECIFIC EXAMPLES

KEY POINTS

✓ *What is the general mechanism of electrophilic aromatic substitution?*

✓ *What are some common examples of electrophilic aromatic substitution?*

The general mechanism for **electrophilic aromatic substitution** (EAS) involves attack of the ring's pi cloud on the electrophile to give a resonance-stabilized intermediate. Loss of a proton from the site of electrophilic attachment rearomatizes the ring.

Several common types of EAS reactions are shown in **Table 1.1**. The mechanisms for these reactions differ mostly in the steps leading up to the formation of the electrophile, E$^+$.

Table 1.1 Survey of Some EAS Reactions			
NAME	**REAGENTS**	**E+ (EFFECTIVE)**	**PRODUCT**
Chlorination	$Cl_2/FeCl_3$	(Cl$^+$)	Ar—Cl
Bromination	$Br_2/FeBr_3$	(Br$^+$)	Ar—Br
Nitration	HNO_3/H_2SO_4	NO_2^+	Ar—NO_2
Sulfonation	H_2SO_4/SO_3	HSO_3^+	Ar—SO_3H
Alkylation	$R—X/AlCl_3$	R$^+$ (R$^+$)	Ar—R
Acylation	$RCOCl/AlCl_3$	RC≡O$^+$	Ar—COR

Topic Test 3: General EAS Reaction and Some Specific Examples

Short Answer

Provide unambiguous structural formulas for the organic products that result from treatment of benzene with each of the following reagents and/or conditions.

1. $Cl_2/FeCl_3$

2. SO_3/H_2SO_4

3. $(CH_3)_3CCl/AlCl_3$

4. Draw all reasonable resonance forms for the intermediate that forms during the bromination of benzene.

Specify the reagents and/or conditions one could use to convert benzene into each of the following.

5. Nitrobenzene

6.

Topic Test 3: Answers

1. Benzene ring with —Cl

2. Benzene ring with —SO$_3$H

3. Benzene ring with —C(CH$_3$)$_3$ group (central C bonded to CH$_3$, CH$_3$, CH$_3$)

4. $$\left[\quad \underset{\oplus}{\text{ring with } H, Cl} \quad \longleftrightarrow \quad \underset{\oplus}{\text{ring with } H, Cl} \quad \longleftrightarrow \quad \underset{\oplus}{\text{ring with } H, Cl} \quad \right]$$

5. HNO$_3$, H$_2$SO$_4$

6. CH$_3$CH$_2\overset{\overset{\text{O}}{\|}}{\text{C}}$Cl, AlCl$_3$

TOPIC 4: SUBSTITUENT EFFECTS IN EAS REACTIONS

KEY POINTS

✓ *Will a given ring substituent speed up or slow down an EAS reaction?*

✓ *What are activating and deactivating groups?*

✓ *What regiochemistry results from an EAS reaction on a substituted benzene?*

✓ *What are ortho/para and meta-directing groups?*

✓ *How can one predict the effect of a particular substituent?*

A substituent on an aromatic ring can render that ring more or less reactive than benzene in EAS reactions. Those substituents that make the ring more reactive are called **activating groups** and those that make the ring less reactive are called **deactivating groups**. A substituent can also influence where an electrophile will attach to the ring. Some groups are called **ortho/para directors** because they tend to direct the electrophile to those positions. Other groups are known as **meta directors** because their presence results in predominately *meta* substitution. The activating/deactivating and o/p,m-directing characteristics are mostly predictable and can be generalized with a few rules.

Rule 1. Activating groups are usually also *o/p* directors and have at least one nonbonded electron pair on the atom attached directly to the ring.

$$-\ddot{N}H_2 \qquad -\ddot{N}R_2 \qquad -\ddot{O}H \qquad -\ddot{O}R \qquad -\ddot{O}\overset{\overset{\text{O}}{\|}}{C}R \qquad -\overset{\overset{\text{O}}{\|}}{\underset{\underset{H}{}}{\ddot{N}}}CR$$

Rule 2. Deactivating groups are usually also *meta* directors and have the general form —W=Y, where an unsaturation (usually a pi bond) is in conjugation with the ring and Y is an electronegative atom.

$-NO_2$ $-SO_3R$ $-CO_2H$ $-COR$ $-CONR_2$ $-CN$

Rule 3. Alkyl groups are mildly activating and are *o/p* directors but they do not fit the general form described in rule 1 above.

$-CH_3$ $-CH_2CH_3$ $-CH(CH_3)_2$ $-C(CH_3)_3$

Rule 4. Halogens and nitroso are mixed effect substituents in that they are mildly deactivating but favor *ortho/para* substitution rather than *meta*.

$-Cl$ $-Br$ $-I$ $-N{=}O$

Topic Test 4: Substituent Effects in EAS Reactions

True/False

1. Monobromination of aniline yields mostly *meta*-bromoaniline.

2. Toluene undergoes most EAS reactions faster than benzaldehyde does.

Multiple Choice

3. Which compound below is likely to nitrate at the *meta* position?

   a.   b.   c.

   d. All of the above
   e. None of the above

4. Which compound below will likely undergo acylation the fastest?
   a. Benzene
   b. Ethylbenzene

   c.   d.   e.

Short Answer

5–6. Show the structure(s) expected for the monosubstitution products of each reaction below.

5.

6.

Topic Test 4: Answers

1. **False.** The NH$_2$ group on the aromatic ring of aniline is an *ortho/para* director so the monobromination reaction will likely yield a mixture of *ortho-* and *para*-bromoaniline.

2. **True.** The deactivating carbonyl conjugated with the ring in benzaldehyde slows down any EAS reaction. The mildly activating methyl group of toluene enhances the reactivity of that aromatic ring toward EAS.

3. **c.** Nitrobenzene bears a *meta*-directing substituent. The product of further nitration will likely be *m*-dinitrobenzene. Compounds shown in responses a and b both have lone pairs conjugated with the aromatic ring. Substituents of this form are o/p directors.

4. **e.** PhOCH(CH$_3$)$_2$ contains the most activated aromatic ring. Ethyl benzene is only mildly activated, and the compounds pictured in c and d are deactivated.

5. The aromatic ring bears two substituents (they happened to be connected to one another but no matter). An incoming electrophile will preferentially attach to a position *ortho* or *para* to the o/p-directing alkyl group and *meta* to the *meta*-directing nitro group.

6. The two rings of this starting material differ in reactivity. The ring bearing the nitro group is deactivated relative to the other ring that has only an alkyl substituent that is a mildly activating o/p director. *Para* substitution will likely predominate for steric reasons, but some *ortho* is also expected on that ring.

TOPIC 5: NUCLEOPHILIC AROMATIC SUBSTITUTION

KEY POINTS

✓ *What products come from reactions of nucleophiles with aromatic compounds?*

✓ *What mechanism(s) are possible for nucleophilic aromatic substitution?*

✓ *What is benzyne?*

Some substitution reactions on aromatic substrates occur by other (non-EAS) mechanisms. If the starting material bears a suitable leaving group, nucleophilic substitution can take place by one of two mechanisms. Aromatics that bear one or more electron-withdrawing groups (deactivating groups for EAS) in positions *ortho/para* to the leaving group favor an **addition-elimination** mechanism.

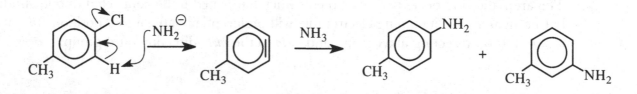

Those aromatics that bear a leaving group but do not have electron-withdrawing substituents can react under more vigorous conditions via an **elimination-addition** pathway known as the **benzyne** mechanism. Note that the regiochemistry is sometimes uncertain for benzyne reactions.

Topic Test 5: Nucleophilic Aromatic Substitution

Multiple Choice

1. Which of the following is most likely to undergo a nucleophilic aromatic substitution reaction via the addition-elimination mechanism?
   a. Chlorobenzene
   b. *meta*-Dinitrobenzene
   c. 1-Chloro-2,4-dinitrobenzene
   d. All of the above
   e. None of the above

2. Which of the following will likely react with KOH/H_2O at elevated temperature and pressure to yield a single organic product via the benzyne pathway?
   a. Chlorobenzene
   b. *p*-Chlorotoluene

c. *m*-Chlorotoluene
d. *o*-Chlorotoluene
e. All of the above

Short Answer

3. What reagents and conditions might be used to covert bromobenzene into aniline?

4–6. Provide unambiguous structural formulas for the missing organic products.

4. 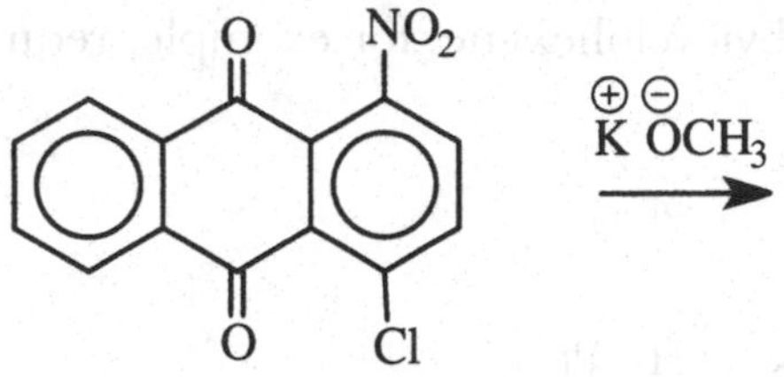
$$\overset{\oplus\ \ominus}{K\ OCH_3} \longrightarrow$$

5. 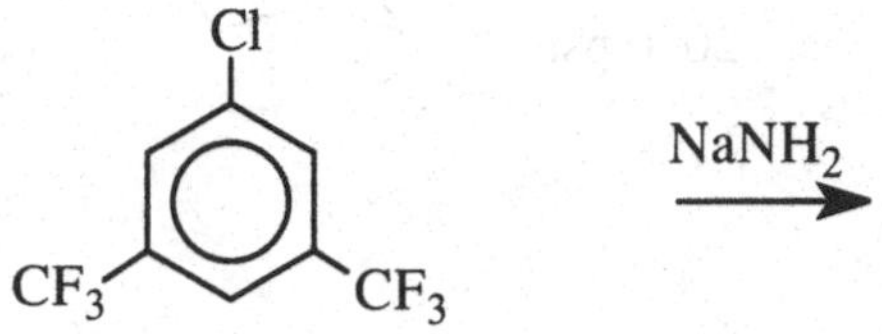
$$NaNH_2 \longrightarrow$$

6. 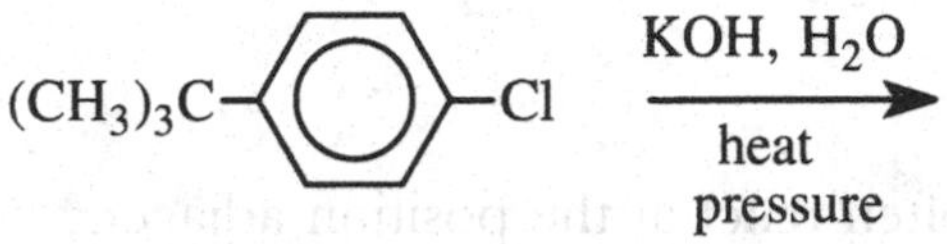 $(CH_3)_3C$—⟨ ⟩—Cl $\xrightarrow[\text{pressure}]{\text{KOH, H}_2\text{O}\ \ \text{heat}}$

Topic Test 5: Answers

1. **c.** For the addition-elimination reaction mechanism to be feasible, the aromatic ring must bear a leaving group (chloride in this case) and electron-withdrawing groups on the ring that will stabilize the anionic intermediate.

2. **a.** The three chlorotoluene isomers would each yield more than one product via the benzyne mechanism

3. $NaNH_2/NH_3$ with heat and pressure (benzyne mechanism).

4.

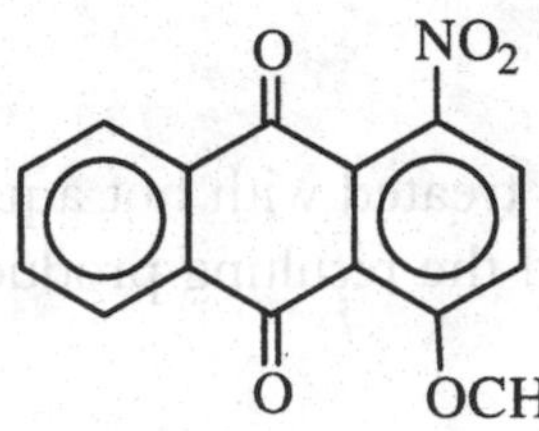

5.

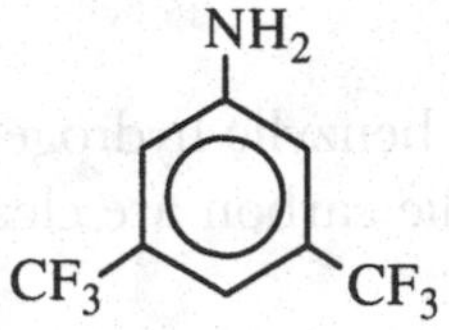

6. 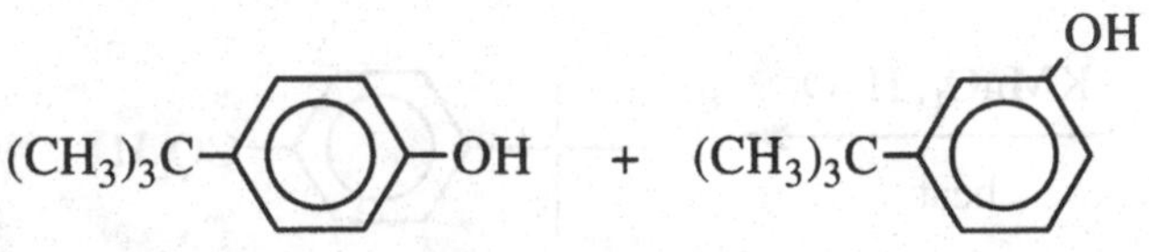 $(CH_3)_3C$—⟨ ⟩—OH + $(CH_3)_3C$—⟨ ⟩ (OH)

TOPIC 6: OTHER REACTIONS OF AROMATIC COMPOUNDS

KEY POINTS

✓ *How can aromatic rings be reduced (hydrogenated)?*

✓ *What is benzylic bromination and what reagents will cause it?*

✓ *How can alkyl substituents on aromatic rings be oxidized?*

The pi systems of benzene and its derivatives are unreactive to the hydrogenation reactions used for alkenes and alkynes. Reduction of toluene to methylcyclohexane, for example, requires the use of a powerful catalyst and/or high pressure.

The alkyl side chains attached to aromatic rings often react at the position adjacent to the ring (called the **benzylic** position). Halogenation or oxidation at the benzylic position is generally possible provided there is at least one benzylic hydrogen atom in the starting material. Chlorination and bromination can be carried out under free radical conditions with molecular halogen or *N*-bromosuccinimide (NBS).

When alkyl groups with one or more benzylic hydrogen atoms are treated with hot aqueous permanganate, all carbons except benzylic carbon are cleaved off, and the resulting product is an aromatic carboxylic acid.

Topic Test 6: Other Reactions of Aromatic Compounds

True/False

1. Benzylic hydrogens are attached directly to the aromatic ring.

2. t-Butylbenzene can be oxidized to benzoic acid with $KMnO_4/H_2O/$heat.

3. The hydrogenation conditions used to reduce aromatic rings are usually more vigorous than those used for the hydrogenation of alkenes or alkynes.

Multiple Choice

4. What reagents and conditions could be used to convert *trans*-1,2-diphenylethene into 1,2-dicyclohexylethane?
   a. Excess H_2/Pd
   b. Excess H_2/Pt, pressure
   c. $KMnO_4/H_2O/$heat
   d. All of the above
   e. None of the above

Short Answer

5–6. Complete the following reactions with unambiguous structural formulas for the major organic products.

5.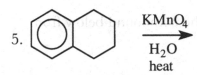
   KMnO$_4$
   $\xrightarrow{\hspace{1cm}}$
   H$_2$O
   heat

6.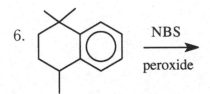
   NBS
   $\xrightarrow{\hspace{1cm}}$
   peroxide

Topic Test 6: Answers

1. **False.** Benzylic hydrogens are those on the carbon attached directly to the ring.

2. **False.** The t-butyl group bears no benzylic hydrogens and therefore the permanganate cannot "bite into" the molecule. Recall that oxidations of alkyl side chains require benzylic hydrogens.

3. **True.** Alkenes and alkynes will usually hydrogenate at room temperature with a Pd catalyst, but aromatic rings normally require Pt or Rh catalysis and high pressure.

4. **b.** These hydrogenation conditions are strong enough to reduce the aromatic rings and therefore will also reduce the alkene.

5. [structure: benzene ring with two CO₂H groups ortho to each other]

6. [structure: bicyclic ring with gem-dimethyl groups and Br]

DEMONSTRATION PROBLEM

Show the reagents and/or conditions one could use to prepare the compound below from toluene. More than one step may be required.

[structure: benzene ring with Cl and CO₂H substituents]

Solution

The carboxylic acid group in the product must have resulted from oxidation of the methyl on toluene using $KMnO_4$ and H_3O^+. The bromine was placed on the ring using Cl_2 and $FeCl_3$. Note that the order these reagents are applied is critical because different regiochemistry results if the steps are reversed.

[reaction scheme row 1: toluene (CH₃) → $\xrightarrow[\text{H}_2\text{O, heat}]{\text{KMnO}_4}$ → benzoic acid (CO₂H) → $\xrightarrow[\text{FeCl}_3]{\text{Cl}_2}$ → meta-chlorobenzoic acid (Cl, CO₂H)]

[reaction scheme row 2: toluene (CH₃) → $\xrightarrow[\text{FeCl}_3]{\text{Cl}_2}$ → para-chlorotoluene (CH₃, Cl) → $\xrightarrow[\text{H}_2\text{O, heat}]{\text{KMnO}_4}$ → para-chlorobenzoic acid (CO₂H, Cl) (+ ortho)]

Chapter Test

True/False

1. 1,3-Cyclohexadiene is aromatic.

2. Phenyl is an aromatic alcohol.

3. Aromatic compounds do not generally undergo addition reactions.

4. The formula for styrene is C_8H_8.

5. Benzene has no benzylic hydrogens.

Multiple Choice

6. Which of the following is a requirement for a compound to be aromatic?
   a. A strong odor
   b. Complete conjugation around a ring
   c. Must be benzene or a substituted benzene
   d. All of the above
   e. None of the above

7. Which step(s) below will convert toluene to *m*-chlorobenzoic acid?
   a. Water and Cl_2
   b. $Cl_2/FeCl_3$ followed by hot aqueous permanganate
   c. Hot aqueous permanganate and then $Cl_2/FeCl_3$
   d. All of the above
   e. None of the above

8. What step(s) below will convert benzene to *p*-bromobenzenesulfonic acid?
   a. H_2SO_4/HNO_3 and then $Br_2/FeBr_3$
   b. $Br_2/FeBr_3$ and then H_2SO_4/HNO_3
   c. H_2SO_4/SO_3 and then $Br_2/FeBr_3$
   d. $Br_2/FeBr_3$ and then H_2SO_4/SO_3
   e. None of the above

9. Which compound below will undergo EAS reactions the fastest?

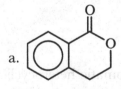
a.

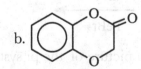
b.

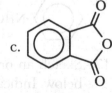
c.

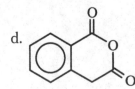
d.

e.

10. The most likely product from the reaction of toluene with NBS and peroxide is
    a. *ortho*-bromotoluene
    b. *meta*-bromotoluene

c. *para*-bromotoluene

d. 2,4,6-tribromotoluene

e. benzyl bromide (PhCH₂Br)

Short Answer

11–16. Provide unambiguous structural formulas for the missing organic products.

11.

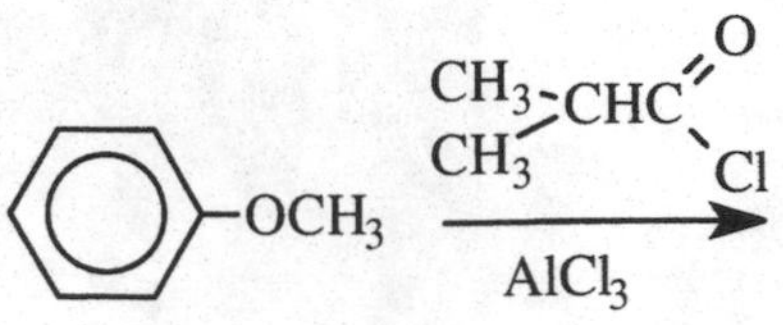

12.

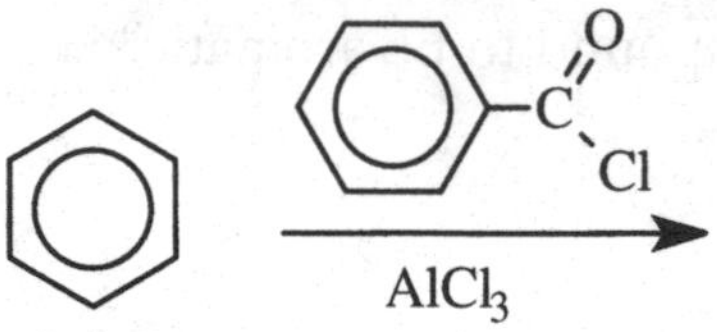

13.

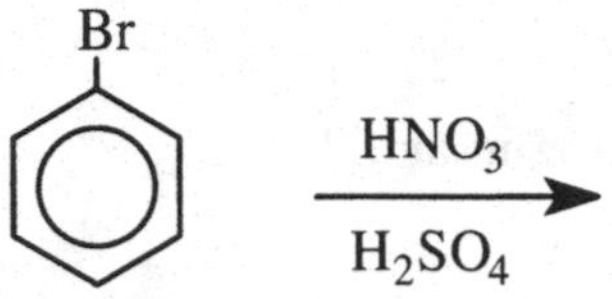

14.

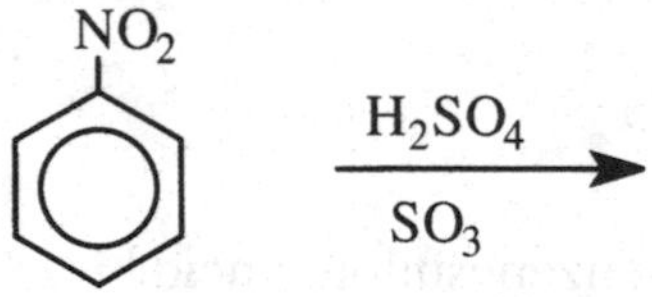

15.

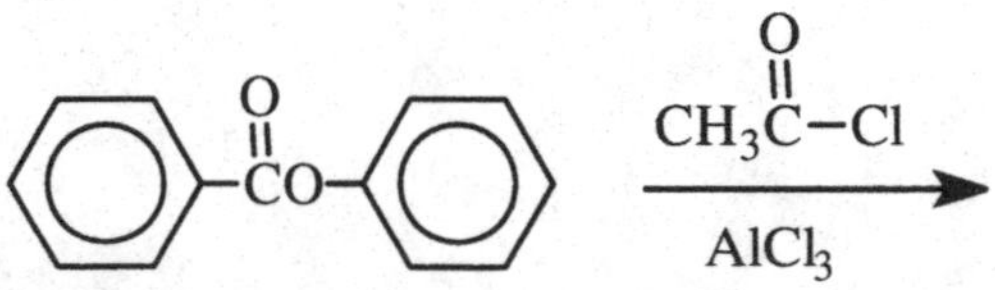

16.

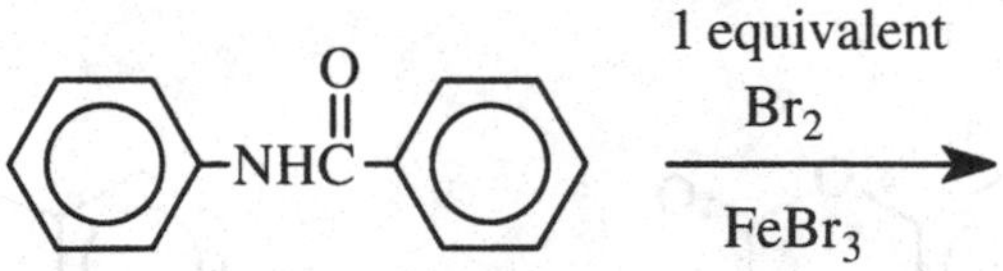

17. Sketch an orbital overlap picture of the pi system for the aromatic heterocyclic compound below. Indicate the positions of all pi and nonbonded electrons and specify the hybridization of all ring atoms.

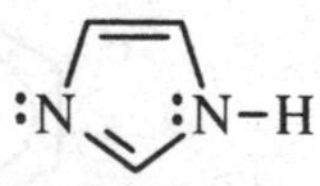

18. The *ortho*- and *para*-bromonitrobenzene each undergo nucleophilic aromatic substitution reactions with hydroxide at 130°C, yet *meta*-bromonitrobenzene is inert to these conditions. Explain this observation. (Hint: Consider resonance and the key mechanistic intermediates in each case.)

19. Show how one could synthesize *p*-nitrobenzoic acid from benzene or toluene and any other needed reagents. More than one step may be required.

20. Show the reagents and/or conditions one could use to carry out the transformation below. More than one step may be required.

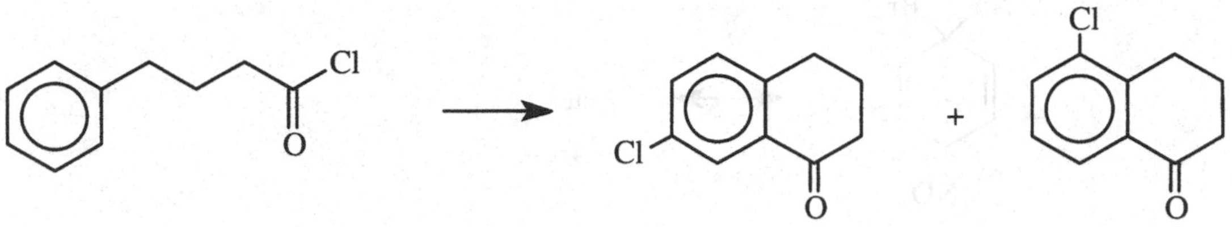

Chapter Test: Answers

1. **F** 2. **F** 3. **T** 4. **T** 5. **T** 6. **b** 7. **c** 8. **d** 9. **b** 10. **e**

11. 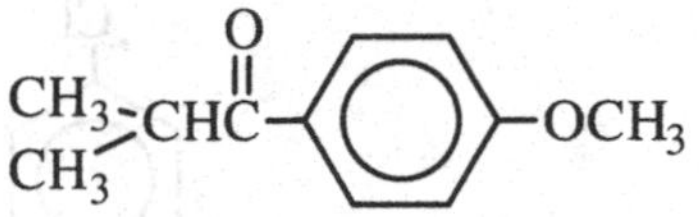 (+ *ortho*)

12.

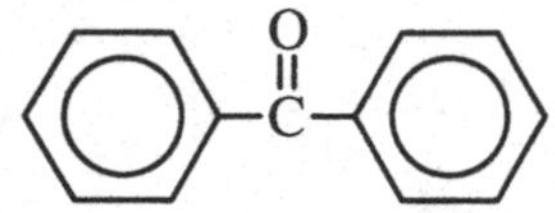

13.  (+ *ortho*)

14.

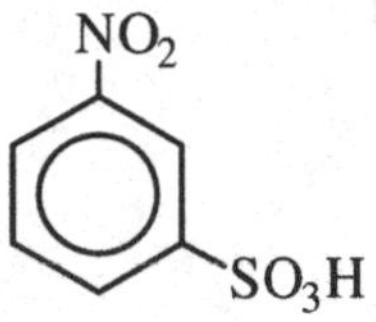

15. 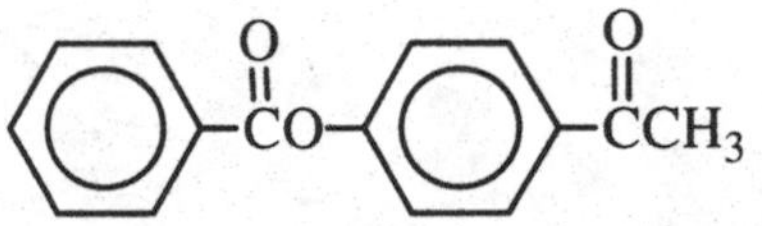 (+ *ortho* on same ring)

16. 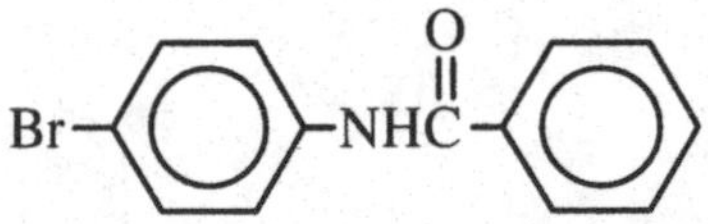 (+ *ortho* on same ring)

17. All ring atoms *sp*² hybridized

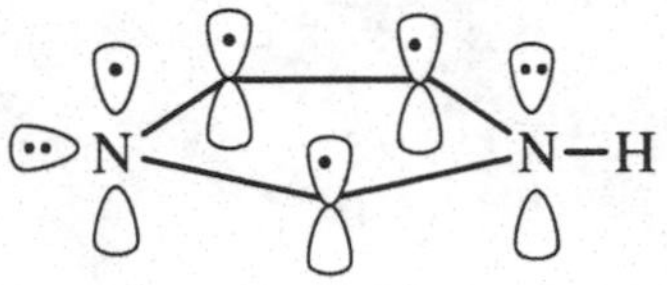

18. The intermediates that result from nucleophilic attack on C(1) are resonance-stabilized anions. The *ortho* and *para* isomers have resonance forms in which the negative charge is stabilized by the strong electron-withdrawing nitro group. The nitro on the intermediate from the *meta* isomer is not able to delocalize the negative charge.

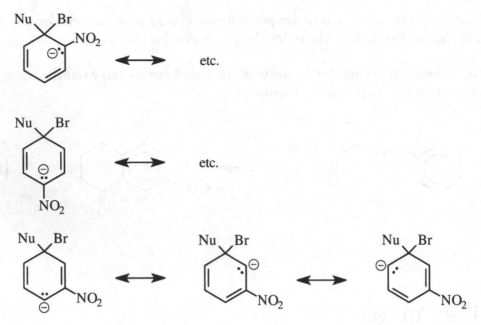

19. Treat toluene with HNO_3/H_2SO_4 and then hot aqueous permanganate.

20. $AlCl_3$ (intramolecular acylation) and chlorinate with $Cl_2/FeCl_3$.

Check Your Performance

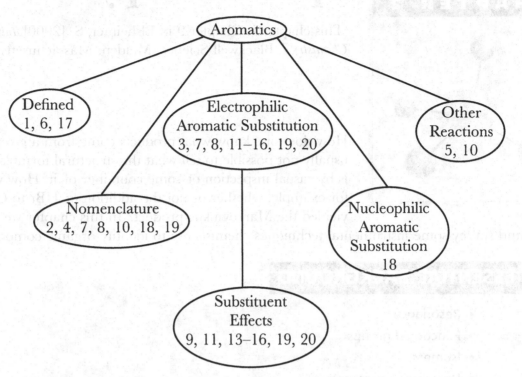

Note the number of questions in each grouping that you got wrong on the chapter test. Identify areas where you need further review and go back to relevant parts of this chapter.

CHAPTER 2

Spectroscopy

(This chapter is Chapter 9 in Elsheimer, S. [2000] *Introduction to Organic Chemistry I*. Blackwell Science, Malden, Massachusetts.)

How does one know what products come from a given reaction? It is usually not possible to tell what the structural formula for a compound is by casual inspection of some container of it. How would one know, for example, whether or not the addition of HBr to $CH_3CH{=}CH_2$ yielded the Markovnikov product? In this chapter we address this issue and survey some instrumental techniques chemists use to identify organic compounds.

ESSENTIAL BACKGROUND

- **Resonance**
- **Functional groups**
- **Isomers**
- **Degree of unsaturation**
- **Relative stability of carbocations**
- **Relative stability of free radicals**

TOPIC 1: MASS SPECTROMETRY

KEY POINTS

✓ *What takes place at the microscopic level when a mass spectrum is measured?*

✓ *What is a molecular ion?*

✓ *What is a base peak in a mass spectrum?*

✓ *What can be learned about a compound by examination of its mass spectrum?*

Most **mass spectra** are acquired using an instrument that places a small amount of gaseous sample in an evacuated chamber where it is bombarded by high-energy electrons. If one of these electrons strikes a molecule of sample with sufficient energy, one of the electrons from the sample molecule is dislodged, leaving behind a positively charged odd-electron species called a **radical cation**. Because the mass of the ejected electron is negligible, the mass and molecular formula for this radical cation are essentially the same as the original sample molecule. This radical cation is usually designated as M^+ and called the **molecular ion**. The charged particles are accelerated by a magnetic field toward a detector that records the impact and registers the mass. The detection equipment of a mass spectrometer is designed to detect ions but not uncharged species. A molecular ion may be so unstable that it decomposes before being detected. When a radical cation fragments, the resultant particles will be radicals and cations. Favorable fragmentation will result in the most stable cations and radicals. The intensity of a peak in the mass spectrum is proportional to the number of ions of that mass reaching the detector. The most intense (tallest) peak of a mass spectrum is called the **base peak**. In some cases the molecular ion is also the base peak. A mass spectrum usually allows one to determine the molecular mass of a compound (and therefore possible molecular formulas) and fragmentation pathways. Examining the spectrum for intensities and masses of molecular and fragment ions can help identify some molecules.

Topic Test 1: Mass Spectroscopy

True/False

1. The tallest (most intense) peak in a mass spectrum is the molecular ion.

2. The molecular ion for benzene will likely appear in the mass spectrum at $m/z = 78$.

3. The base peak of a mass spectrum is usually abbreviated as M^+.

4. Compounds with the same molecular formula will also produce the same mass spectrum.

Multiple Choice

5. Which pair below will most likely give the same molecular ion?
   a. Hexane and cyclohexane
   b. Cyclohexane and cyclohexene
   c. Cyclohexane and *trans*-2-hexene
   d. All of the above
   e. None of the above

6. Which compound will most likely show a strong peak at M-15 in its mass spectrum?
   a. Cycloheptane
   b. 1,1-Dimethylcyclopentane
   c. Biphenyl (Ph-Ph)
   d. All of the above
   e. None of the above

Short Answer

7. The mass spectrum of an unidentified hydrocarbon shows a $M^+ = 80$. Determine the formula and propose a possible structure.

8. Which compound below is most likely to show a large mass spectral peak at $m/z = 69$? Explain your choice.

$$(CH_3)_2C{=}CHCH_2CH_3 \qquad (CH_3)_2C{=}C(CH_3)_2$$

Topic Test 1: Answers

1. **False.** The most intense peak in a mass spectrum is called the base peak. Although the molecular ion is sometimes also the base peak, that is not always the case.

2. **True.** The formula for benzene is C_6H_6 with a molecular weight (MW) of 78.

3. **False.** The abbreviation M^+ stands for the molecular ion.

4. **False.** Isomers will have the same M^+ but can, and often do, fragment differently and therefore have different mass spectra.

5. **c.** To have the same M^+ the compounds must be isomers. Both these compounds have the formula C_6H_{12}.

6. **b.** This is the only compound listed that has methyl appendages. If one of the methyl groups is lost from the M^+, the resulting fragment ion will have a mass that is 15 amu lighter and will be tertiary.

7. A hydrocarbon with an MW = 80 must be C_6H_8. Any compound with that formula is a possible correct answer: 1,3-cyclohexadiene or 1,3,5-hexatriene, etc.

8. $(CH_3)_2C{=}CHCH_2CH_3$ is most likely to show m/z = 69. Both compounds have MW = 84. A peak at 69 corresponds to loss of methyl (M-15). Loss of methyl from the molecular ion of $(CH_3)_2C{=}C(CH_3)_2$ would yield the relatively unstable vinylic carbocation. Loss of methyl from the molecular ion of $(CH_3)_2C{=}CHCH_2CH_3$ (i.e., methyl that is carbon 5 in 2-methyl-2-pentene) produces a resonance-stabilized allylic carbocation.

TOPIC 2: ULTRAVIOLET VISIBLE

KEY POINTS

✓ *What occurs when a molecule absorbs ultraviolet-visible (UV-vis) light?*

✓ *What kind of molecules will absorb UV-vis light?*

✓ *What can be learned about a compound by examining its UV-vis spectrum?*

UV-vis spectroscopy measures the transition of electrons from occupied to unoccupied molecular orbitals. The amount of energy required for this transition depends on the energy gap between the two orbitals. The kinds of transitions that fall within the usual UV-vis range are n $\rightarrow$ π^* and $\pi \rightarrow \pi^*$. Conjugation lowers the ΔE between orbitals, and normally compounds that absorb UV-vis light of longer wavelengths than 210 nm have such conjugation. In general, more conjugation causes a longer wavelength of absorption. Conjugation can include both pi bonds and non-bonded electron pairs. Systems with similar conjugation tend to have similar UV-vis spectral characteristics. A **wavelength of maximum absorption** is designated by the symbol λ_{max}. Some representative UV-vis data are shown in **Table 2.1**.

Table 2.1 Some Ultraviolet-Visible Absorption Maxima	
COMPOUND	**λ_{MAX} (NM)**
R—CH=CH—R	~165
1,3-butadiene	217
1,3,5-hexatriene	258
1,3,5,7-octatetraene	290
2-methyl 1,3-butadiene	220
3-buten-2-one	219
Benzene	204, 254
Phenol	210, 270

Topic Test 2: Ultraviolet Visible

True/False

1. Cyclohexene absorbs UV-vis light at a wavelength longer than 210 nm.

2. *trans*-1,3-Pentadiene has a longer λ_{max} than 1,4-pentadiene does.

3. Methyl vinyl ether ($CH_3OCH{=}CH_2$) has a longer λ_{max} than $CH_3CH{=}CH_2$ does.

Multiple Choice

4. Which compound below will have the longest λ_{max}?
   a. Cyclohexane
   b. Cyclohexene
   c. Benzene
   d. Toluene
   e. Nitrobenzene

5. Which of the following is *not* an ultraviolet transition normally observed above 210 nm?
   a. n-π*
   b. π-π*
   c. σ-σ*
   d. All of the above
   e. None of the above

Short Answer

6. An unidentified compound with the formula C_6H_8 was found to have no UV-vis absorption above 200 nm. Propose a possible structure.

Topic Test 2: Answers

1. **False.** Cyclohexene has only a single pi bond and is not conjugated.

2. **True.** 1,3-Pentadiene is conjugated and 1,4-pentadiene is not.

3. **True.** The lone pairs of electrons on the oxygen are part of the extended pi system.

4. **e.** Besides the aromatic ring, the nitro group is also part of the conjugation.

5. **c.** This transition will only occur at very high energies that correspond to wavelengths much shorter than 210 nm.

6. There are many possible correct answers. Any C_6 hydrocarbon with three degrees of unsaturation and no conjugation is a reasonable choice (e.g., 1,4-cyclohexadiene or 3-vinylcyclobutene or cyclobutylethyne, etc.).

TOPIC 3: INFRARED SPECTROSCOPY

KEY POINTS

✓ *What occurs when a molecule absorbs infrared (IR) light?*

✓ *What can be learned about a compound from its IR spectrum?*

✓ *What are some important regions in the IR spectrum?*

To a first approximation, bonds connecting atoms can be viewed as springs connecting two masses. These "springs" bend and stretch with energies that depend on the masses of the attached atoms or groups and the inherent strength of the spring. The stretching and bending frequencies of the bonds are expressed in wavenumbers (cm^{-1}) and can be measured with IR spectroscopy. A typical IR spectrum extends from about $4000 \, cm^{-1}$ at left to about $400 \, cm^{-1}$ at right. When a molecule absorbs IR energy, the bonds are set into vibration. The precise position of the absorption frequency within the spectrum provides information on what "kinds" of bonds or functional groups are present. Some general regions of the spectrum and the types of vibrations to which they correspond are represented below along with a list of some common IR peak positions.

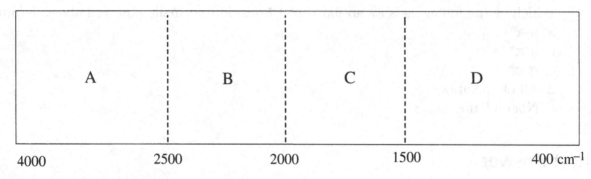

A: Stretching of single bonds to hydrogen (C—H, O—H, N—H)

B: Triple-bond stretch (C≡C, C≡N)

C: Double-bond stretch (C=C, C=O, C=N)

D: "Fingerprint region," stretching single bonds not to H, bond bending, harmonics.

Many of the most common and diagnostic bonds are shown in **Table 2.2**.

Table 2.2 Some Common IR Absorption Bands	
FUNCTIONAL GROUP	**APPROXIMATE BAND POSITION (cm^{-1})**
C—H	2850–2960
=C—H	3020–3100
C≡C—H	3300
Aromatic-H	3030
O—H	3400–3640 (alcohol), 2500–3100 (carboxylic acid)
N—H	3310–3500
C≡C	2100–2260
C≡N	2210–2260
C=O	1670–1780
NO_2	near 1350 and 1530
Aromatic C—C	near 1500 and 1600
C—O	1050–1150
C—Cl	600–800
C—Br	500–600
C—I	500

Topic Test 3: Infrared Spectroscopy

True/False

1. Cyclohexane and 1-hexene are isomers and therefore have nearly identical IR spectra.

2. Cyclohexane and cycloheptane are not isomers, yet they have nearly identical IR spectra.

3. Most bond bending occurs with energies that correspond to the fingerprint region.

Multiple Choice

4. The region of the IR spectrum where double bonds stretch is
   a. $4000–2500\,cm^{-1}$.
   b. $2500–2000\,cm^{-1}$.
   c. $2000–1500\,cm^{-1}$.
   d. the fingerprint region.
   e. None of the above

5. Which class of compounds below shows a strong IR absorption near $1700\,cm^{-1}$.
   a. Alcohol
   b. Ether
   c. Alkyne
   d. Aldehyde
   e. None of the above

Short Answer

6. Briefly explain how IR spectroscopy could distinguish between 1-pentyne and 2-pentyne.

7. Briefly explain how IR spectroscopy could distinguish between cyclohexane and benzene.

Topic Test 3: Answers

1. **False.** Although they are isomers, these compounds have different kinds of bonds and will therefore show different peaks in their IR spectra.

2. **True.** Both these compounds are medium-sized cyclic alkanes with similar bonds. They will therefore have similar absorption bands in their IR spectra.

3. **True.** The region to the right of 1500 wavenumbers (the fingerprint region) is where the bond-bending energies are found.

4. **c**

5. **d.** This is the only carbonyl-containing class in the list, and a peak near 1700 wavenumbers is likely a carbonyl.

6. The terminal alkyne 1-pentyne has a hydrogen atom bound directly to an sp-hybridized carbon. That bond stretches near $3300\,cm^{-1}$. The internal alkyne 2-pentyne will not have an IR signal in this region.

7. Cyclohexane is a cycloalkane that will show sp^3 C—H stretch in the 2900-cm^{-1} region. Benzene has only aromatic C—H stretches that appear near 3030 cm^{-1}. Benzene will also show the aromatic C—C stretches near 1500 and 1600 cm^{-1}.

TOPIC 4: NUCLEAR MAGNETIC RESONANCE AND ^{1}H-NMR

KEY POINTS

✓ *What kinds of nuclei does nuclear magnetic resonance (NMR) detect?*

✓ *What occurs when NMR is being measured?*

✓ *How does proton NMR reveal the number of different hydrogen environments?*

✓ *What is meant by "chemical shift" and what does it reveal?*

✓ *How can one determine the number of hydrogens in each environment?*

✓ *What information is contained in ^{1}H-NMR peaks that appear as multiplets?*

The **nuclear magnetic resonance** (NMR) phenomenon involves certain nuclei that can be "flipped" between energy states by radio frequency in the presence of an applied magnetic field. Nuclei of atoms that have an odd atomic number and/or odd atomic mass have a nuclear "spin." Among such nuclei are ^{1}H, ^{13}C, ^{15}N, ^{19}F, and ^{31}P. Spinning nuclei can be aligned with or against an applied magnetic field. Alignment with the applied magnetic field (parallel) is a lower energy state than alignment against the applied magnetic field (antiparallel). Nuclei can be stimulated to flip from the lower to the higher energy state when the correct amount of energy is applied. When that occurs, the nucleus is said to be "in resonance." The absorption of energy to bring a nucleus into resonance can be recorded in the form of an NMR spectrum. The energy required to induce this nuclear resonance corresponds to the long wavelengths and is normally in the range of radio waves. The precise energy required will depend on several factors, including the type of nucleus and the environment of that nucleus.

The most common kind of NMR spectroscopy is proton NMR (^{1}H-NMR). In general, a ^{1}H-NMR spectrum will help to answer four important questions about the structure of the compound:

Question 1: How many different ^{1}H environments are present? This is found through peak counting. In theory there is a different peak for each chemically different hydrogen environment within a molecule. Accidental superimposition of peaks is common so the actual number of peaks observed is often fewer than the number of ^{1}H environments.

Question 2: What is the nature of each hydrogen environment? The "chemical shift" or the position of the peak within the spectrum indicates this. Proton NMR shifts are reported in **parts per million** (ppm or δ) **downfield** (to the left) of teramethylsilane (**TMS**), which is assigned the value of zero. Generally, protons near electronegative substituents are shifted downfield to a larger extent. Sometimes proton shifts are affected strongly by the presence of multiple bonds or aromatic rings. Some rough ranges within which specific hydrogen types can be expected are listed in **Table 2.3**.

Table 2.3 Some Proton NMR Chemical Shift Ranges

PROTON ENVIRONMENT AND GENERAL STRUCTURE		APPROXIMATE CHEMICAL SHIFT (δ)
Alkane-like		0–1.5
On sp^3-hybridized carbons adjacent to unsaturation (Bezylic, Allylic, alpha to C=O)	=C-C-H	1.5–2.5
On carbon bearing electronegative atom (Z = halogen, O, N)	—CHZ—	2.5–4.5
Vinylic	C=C—H	4.5–6.5
Aromatic	Ar—H	6.5–8.0
Aldehydic	RCH=O	near 10
Carboxylic acid	RCO_2H	11–13

Question 3: How many protons of each type are present? This information appears in relative terms and is given by the relative areas under the peaks. Most NMR spectrometers are equipped with an integrator that converts peak areas into digitized numbers or altitudes for comparison. The relative areas of the peaks reflect the relative numbers of protons producing them.

Question 4: How many neighboring protons are adjacent to each proton type present? This is deduced from the **splitting pattern** caused by the **coupling** of nearby protons. In general, a proton NMR signal coupled to n equivalent nuclei will be split into n + 1 lines. If the neighbors are not strictly equivalent but coincidentally couple by similar amounts, the n + 1 relationship can still hold, but this is not always the case. More complex and non-first-order coupling patterns sometimes appear, especially when the differences between the chemical shifts of the nuclei are not much larger than the magnitude of the coupling between them. Some common coupling situations and patterns are given in **Table 2.4**.

Table 2.4 Some Common Bonding Arrangements and Proton NMR Coupling Patterns

MULTIPLET NAME	RELATIVE LINE AREAS	POSSIBLE STRUCTURES (G AND X BEAR NO H)
singlet		G—CH_3
doublet	1:1	CH_3—CHX_2 $(CH_3)_2CH$—G
triplet	1:2:1	CH_3—CH_2—G
quartet	1:3:3:1	CH_3—CH_2—G
quintet	1:4:6:4:1	(X—CH_2)$_2$CH—G
septet	1:6:15:20:15:6:1	$(CH_3)_2CH$—G

There are several common patterns of shift, integration, and splitting that you should recognize, including isolated methyl, ethyl, isopropyl, *t*-butyl, *p*-aromatic, and monosubstituted aromatic (phenyl). These are illustrated in the problems that follow.

Topic Test 4: Nuclear Magnetic Resonance and ^{1}H-NMR

True/False

1. The peaks in the proton NMR spectrum of bromoethane integrate in the ratio 2:3.

2. The peaks in the proton NMR spectrum of $CH_3CH_2OCH_2CH_3$ integrate in the ratio 2:3.

3. 2-Bromo-2-methylpropane shows only a singlet in its proton NMR spectrum.

4. The proton NMR spectrum of ethane appears as a quartet because each methyl group has three neighboring protons.

Multiple Choice

5. Which of the following can be examined directly by NMR?
   a. ^{12}C
   b. ^{16}O
   c. ^{4}He
   d. All of the above
   e. None of the above

6. Which isomer below would you expect to show a proton NMR signal near 7 ppm that integrates for five protons?
   a. p-Xylene (p-dimethylbenzene)
   b. m-Xylene
   c. o-Xylene
   d. Ethylbenzene
   e. All of the above

Short Answer

7–9. Assume you have an infinitely well-resolved proton NMR spectrum for each of the following. How many environments would you expect to see for each?

7.

8. CH_3CH_2—⬡—NO_2

9. CH_3—⬡—$C=C$ with H, H, H

10. Propose a structure for a compound that has the formula $C_4H_{10}O$ and shows the proton NMR spectrum tabulated below.

δ	Multiplicity	Relative Integral
3.55	Septet	1H
3.30	Singlet	3H
1.2	Doublet	6H

11. Propose a structure for a compound that has the formula $C_{10}H_{14}$ and shows the proton NMR spectrum tabulated below.

δ	Multiplicity	Relative Integral
7.1	Broad singlet	5H
1.2	Singlet	9H

Topic Test 4: Answers

1. **True.** There are two proton environments in CH_3CH_2Br. There are two of one type and three of the other.

2. **True.** There are two proton environments in $(CH_3CH_2)_2O$. There are four of one type and six of the other, but because the integrator can only express the ratios (not absolute numbers) this will appear as $2:3$.

3. **True.** All the protons of $(CH_3)_3CBr$ are equivalent and are not coupled or split.

4. **False.** All six hydrogens of ethane are equivalent and do not couple to one another. The proton NMR spectrum of ethane will appear as a singlet.

5. **e.** The three nuclei mentioned have even masses and atomic numbers.

6. **d.** This is the only monosubstituted benzene listed. For the peak near 7 ppm (aromatic) to integrate for 5H, there can only be one substituent n the ring.

7. Four

8. Four

9. Eight

10. The 3H singlet is an isolated methyl. The 1H septet and 6H doublet are coupled to one another and are the characteristic pattern of an isopropyl group. The formula indicates saturation so the only reasonable structure is 2-methoxypropane: $(CH_3)_2CHOCH_3$.

11. The 9H singlet is characteristic of a *t*-butyl group and the 5H signal near 7 ppm indicates a monosubstituted benzene (phenyl). The formula indicates the degree of unsaturation is four. The two fragments account for the entire structure of *t*-butylbenzene: $Ph—C(CH_3)_3$.

TOPIC 5: CARBON-13 NMR

KEY POINTS

✓ *How is carbon-13 NMR like proton NMR?*

✓ *How is carbon-13 NMR different from proton NMR?*

✓ *How does one interpret a carbon-13 NMR spectrum?*

Carbon-13 NMR spectroscopy is similar to proton NMR in that the number of peaks in the spectrum normally corresponds to the number of different carbon environments and the chemical shifts of the carbon signals provide some indication of the nature of each environment. Carbon-13 NMR differs from proton NMR in that integration is normally not done and the magnitude of each resonance signal depends not only on the number of carbons that produced it but on several other factors that make integration unreliable. The chemical shift range of carbon-13 NMR signals is approximately 220 ppm and is expressed as δ (downfield of TMS). This wide range of shifts makes it less likely that accidental superimposition of peaks will occur. **Table 2.5** shows carbon-13 NMR chemical shift ranges.

Table 2.5 Some Approximate Carbon-13 NMR Chemical Shift Ranges (δ)	
C=O	180–220
Aromatic	120–170
Alkene C=C	100–150
Alkyne C≡C	70–90
C—O	30–85
C—N	10–60
C—Cl	40–85
C—Br	30–70
—CH$_2$— (alkane)	20–65
CH$_3$— (alkane)	10–40

Coupling between adjacent carbons is not generally observed. Carbon-13 NMR spectra are usually acquired in a mode that removes the effect of any coupling to the protons attached to the carbons. Spectra run in this **proton-decoupled** mode show each carbon environment as a singlet. It is possible to establish the number of protons on each carbon by running the spectrum in the coupled mode producing a spectrum in which each carbon signal will be slit into an n + 1 multiplet resulting from the attachment of n protons. The number of attached protons is also often determined by more modern NMR techniques.

Topic Test 5: Carbon-13 NMR

True/False

1. The carbon-13 NMR spectrum of 2-methylpropane has only two lines.

2. A proton-decoupled (normal) carbon-13 NMR spectrum shows splitting due to the coupling of adjacent carbons.

Multiple Choice

3. Which isomer below will show a carbon-13 NMR signal between 180 and 220 ppm?

a. ▷—CH$_2$OH   b. □—O   c. (ketone structure)

d. All of the above

e. None of the above

4. The number of protons attached to a given carbon

   a. is not revealed in a proton-decoupled carbon-13 NMR spectrum.

   b. is revealed if the carbon-13 NMR spectrum is run in the proton-coupled mode.

   c. can be detected using modern NMR techniques.

   d. All of the above

   e. None of the above

Short Answer

5–6. Identify the number of different carbon environments in the compounds below.

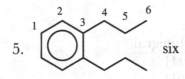

5.

6.

Topic Test 5: Answers

1. **True:** $(CH_3)_3CH$.

2. **False.** There is generally no coupling between adjacent carbons because the natural abundance of carbon-13 is so low (about 1%) and therefore the statistical likelihood that two carbon-13 nuclei will be bounded to one another is small.

3. **c.** A signal between 180 and 220 ppm is likely a carbonyl. The four-carbon ketone is the only carbonyl-containing compound shown.

4. **d**

5.   six   6.   eight

APPLICATION

Organic chemists have been using NMR since the 1960s, but more recently a popular technology based on the NMR phenomenon has been directly applied to the field of medicine. Magnetic resonance imaging provides a noninvasive method for imaging tissues from deep within a patient's body. The patient is placed within the poles of a large magnet and a series of "spectra" are acquired corresponding to cross-sections of all or part of the body. The data are processed by computer and can be displayed as a two-dimensional image or even a three-dimensional virtual reality image.

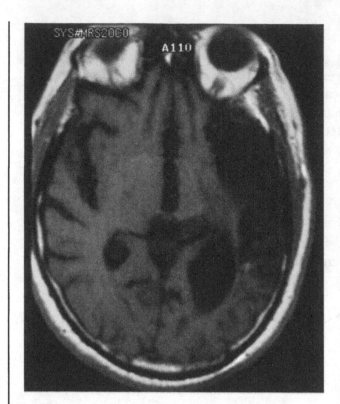

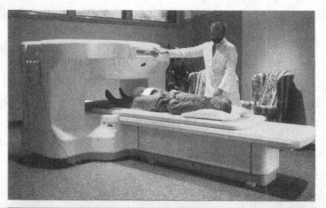

From Mills VM, Cassidy JW, Katz, DI.
Neurologic rehabilitation; a guide to diagnosis,
prognosis, and treatment planning. Malden,
MA: Blackwell Scientific, 1997.

The two figures above are courtesy of the Shields
Health Care Group, Brockton, Massachusetts.

DEMONSTRATION PROBLEM

Deduce a structural formula from spectral data provided.

MS:	$M^+ = 134$		
IR:	Strong peak at $1703\,\text{cm}^{-1}$		
^{1}H-NMR:	ppm	Multiplicity	Integral
	10.0	Singlet	1H
	7.9	Doublet	2H
	7.4	Doublet	2H
	2.7	Quartet	2H
	1.3	Triplet	3H
^{13}C-NMR:	15, 29, 128, 130, 135, 152, and 192 ppm		

Solution

One strategy for solving problems of this type is to summarize what you see easily and then
assemble possible structures from the detected "pieces." The IR peak at $1703\,\text{cm}^{-1}$ is probably a
carbonyl. The singlet at 10 ppm in the proton NMR agrees with the carbonyl above. There are
two recognizable patterns in the proton NMR. The two doublets in the aromatic region that
each integrate for 2H each (or a total of 4H) indicate a *para*-substituted benzene ring. The 2H
triplet and the 3H quartet are the classic combination indicating an ethyl group. Subtracting the

pieces from what we know to be the molecular weight will reveal if all the parts have been identified.

$$
\begin{array}{rl}
134 & M^+ \\
-29 & \text{Aldehyde, O}\!=\!\text{C}\!-\!\text{H} \\
\hline
105 & \\
-76 & p\text{-aromatic, } C_6H_4 \\
\hline
29 & \\
-29 & \text{ethyl, } CH_3CH_2 \\
\hline
0 &
\end{array}
$$

All the mass has been accounted for. The pieces can only attach together in one way, thus giving p-ethylbenzaldehyde.

Now confirm that a proposed structure agrees with all available data. The carbon-13 NMR data were not used to obtain the solution, but they do agree with the proposed structure. The symmetry of the compound leads to only seven carbon environments (even though there are a total of nine carbons). There is a carbonyl (192 ppm), four aromatic signals, and two nonaromatic signals.

Chapter Test

True/False

1. The tallest (most intense) peak in a mass spectrum indicates the compound's MW.

2. A compound that has λ_{max} at longer than 210 nm is probably conjugated.

3. 1,4-Cyclohexadiene has only two peaks in its proton-decoupled carbon-13 NMR spectrum.

4. Protons on aromatic rings normally have a chemical shift near 5 ppm.

5. The protons on C(2) of propane will appear as a septet in the proton NMR spectrum.

6. Methylcyclohexane shows five lines in the carbon-13 NMR spectrum, but cyclohexane shows only one.

Multiple Choice

7. A mass spectral peak at M-17 probably indicates
   a. loss of OH.
   b. loss of Br.
   c. loss of methyl.
   d. Any of the above
   e. None of the above

8. Which compound below has a λ_{max} most like that of 1,3-pentadiene?
   a. $(CH_3)_2CHCH_2CH_2CH_2CH\!=\!CH\!-\!CH\!=\!CH_2$
   b. $CH_2\!=\!CHCH_2CH\!=\!CH_2$

c. $CH_3CH_2CH_2CH{=}CH_2$

d. $CH_3CH_2CH_2CH_2CH_3$

e. All of the above have essentially the same λ_{max}.

Indicate the kind of spectroscopy best associated with each of the following:

9. Stretching and bending of bonds.

10. Electrons are excited to higher energy molecular orbitals.

11. Molecules are bombarded by electrons to yield ions that subsequently fragment.

12. Spinning nuclei exposed to an applied magnetic field are excited to a higher energy state by a radio frequency.

13. Usually reveals the molecular weight of a compound.

14. Indicates the presence of various functional groups.

15. Provides information on the number of different hydrogen or carbon environments.

16. Shows the presence and extent of conjugation.

Combination Spectral Problems

Deduce structural formulas from the spectral data provided.

17. Mass spectrum shows $M^+ = 72$, IR shows a strong peak near $1720\,cm^{-1}$. Carbon-13 NMR shows four lines. The proton NMR is tabulated below.

δ	Multiplicity	Relative Integral
2.4	Quartet	2H
2.1	Singlet	3H
1.1	Triplet	3H

18. $M^+ = 108$; IR 1500, 1600, $1250\,cm^{-1}$; ^{13}C-NMR shows five lines; λ_{max} near 220, 270, and 285 nm

δ	Multiplicity	Relative Integral
6.8–7.2	Multiplet	5H
3.7	Singlet	3H

19. $M^+ = 122$; IR peaks near 1500, 1600 strong broad peak near $3350\,cm^{-1}$; ^{13}C-NMR shows six lines

δ	Multiplicity	Relative Integral
7.15	Multiplet	5H
3.75	Triplet	2H
2.75	Triplet	2H

20. $M^+ = 114$; IR near 2960, 2980, and strong near $1715\,cm^{-1}$; ^{13}C-NMR shows three lines (one is > 200 ppm)

δ	Multiplicity	Relative Integral
3.7	Septet	^1H
1.0	Doublet	6H

21. $M^+ = 117$; $\lambda_{max} \cong 234\,nm$; ^{13}C-NMR shows six lines; IR $2230\,cm^{-1}$

δ	Multiplicity	Relative Integral
7.5	Doublet	2H
7.2	Doublet	2H
2.4	Singlet	3H

Chapter Test: Answers

1. **F** 2. **T** 3. **T** 4. **F** 5. **T** 6. **T** 7. **a** 8. **a**

9. IR

10. UV-vis

11. MS

12. NMR

13. MS

14. IR

15. NMR

16. UV-vis

17.

18. ⬡—OCH₃

19. ⬡—CH₂CH₂OH

20. (structure)

21. CH₃—⬡—C≡N

Check Your Performance

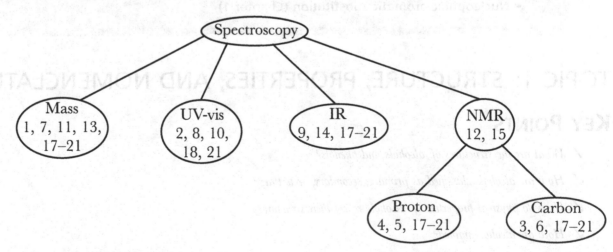

Note the number of questions in each grouping that you got wrong on the chapter test. Identify areas where you need further review and go back to relevant parts of this chapter.

CHAPTER 3

Alcohols, Phenols, and Thiols

Alcohols comprise an important class of organic compounds. They are widely distributed in nature, have broad industrial applications, and are versatile synthetic intermediates. They can be prepared from many different kinds of organic compounds and also serve as the starting materials for numerous organic syntheses. In this chapter we survey the diverse chemistry of alcohols, phenols, and thiols.

ESSENTIAL BACKGROUND

- **Bronsted acids**
- **Resonance**
- **Acid-catalyzed alkene hydration**
- **Oxymercuration of alkenes**
- **Hydroboration of alkenes**
- **Markovnikov's rule**
- **Conversion of alcohols to alkyl halides**
- **Substitution and elimination mechanisms**
- **Zaitsev's rule**
- **Synthesis and properties of Grignard reagents**
- **Electrophilic aromatic substitution (EAS) (Chapter 1)**
- **Electron-withdrawing and electron-donating groups (Chapter 1)**
- **Nucleophilic aromatic substitution (Chapter 1)**

TOPIC 1: STRUCTURE, PROPERTIES, AND NOMENCLATURE

KEY POINTS

✓ *What are the structures of alcohols and phenols?*

✓ *How are alcohols classified as primary, secondary, or tertiary?*

✓ *How do physical properties of alcohols reflect their structure?*

✓ *How are alcohols named?*

Alcohols have an OH bound directly to an sp^3 hybridized carbon. **Phenols** have an OH bound directly to a benzene ring. Although both alcohols and phenols have an OH group bound to carbon, their properties and reactions are sufficiently different that they will be considered separately. Alcohols are classified as primary, secondary, or tertiary depending on whether the alcohol carbon is connected directly to one, two, or three other carbons, respectively.

RCH_2-OH	R_2CH-OH	R_3C-OH	$Ph-OH$
Primary	Secondary	Tertiary	Phenol
$(R \neq H)$			

The polarity of the C–O and O–H bonds makes alcohols and phenols more polar than hydrocarbons or alkyl halides of comparable size and shape. Hydrogen bonding is possible and leads to higher boiling points and melting points due to the strong intermolecular forces. Low-molecular-weight alcohols (about $\leq C_4$) are water soluble, but aqueous solubility decreases as the size of the alkyl portion increases.

The common names for alcohols are two words with the name of the alkyl portion given first followed by the word "alcohol." According to the International Union of Pure and Applied Chemistry (IUPAC), systematic alcohol names are derived from the longest continuous carbon chain that bears the OH. A number is used to designate where along the chain the hydroxyl is located and the suffix "ol" replaces the "e" at the end of the parent alkane name. Any alkyl branches or other substituents are listed alphabetically and assigned numbers as locators with the OH taking priority and receiving the lowest number possible even when that gives higher numbers to any alkyl or halo substituents or any unsaturations. A structure with two or three alcohol groups is called a **diol** or **triol**, but some common names are widely recognized. In some compounds containing multiple functional groups, the OH is designated as a hydroxy substituent on the parent molecule. **Table 3.1** illustrates the use of both common and systematic IUPAC nomenclature.

Table 3.1 Structures and Names of Some Alcohols

STRUCTURE	COMMON NAME	SYSTEMATIC IUPAC NAME
CH_3OH	Methyl alcohol	Methanol
CH_3CH_2OH	Ethyl alcohol	Ethanol
$CH_3CH_2CH_2OH$	Propyl alcohol	1-Propanol
$(CH_3)_2CHOH$	Isopropyl alcohol	2-Propanol
$CH_3CH_2CH_2CH_2OH$	Butyl alcohol	1-Butanol
$(CH_3)_2CHCH_2OH$	Isobutyl alcohol	2-Methyl-1-propanol
$(CH_3)_3COH$	tert-Butyl alcohol	2-Methyl-2-propanol
$CH_2{=}CHCH_2OH$	Allyl alcohol	2-Propen-1-ol
	Benzyl alcohol	Phenylmethanol
	Cyclohexyl alcohol	Cyclohexanol
		4,5-Dibromo-2-methyl-2-hexanol
$HOCH_2CH_2OH$	Ethylene glycol	1,2-Ethanediol
	Glycerol (glycerine)	1,2,3-Propanetriol

Topic Test 1: Structure, Properties, and Nomenclature

True/False

1. Cyclohexanol is a tertiary alcohol.

2. 2-Pentanol is more soluble in water than either 2-methylpentane or 2-chloropentane.

Multiple Choice

3. Which of the following has the highest boiling point?
   a. 1-Propanol
   b. Isopropyl alcohol
   c. Propane
   d. Glycerol (1,2,3-propanetriol)
   e. Ethanol

4. Which of the following is a primary alcohol?
   a. 2-Pentanol
   b. 3-Pentanol
   c. Benzyl alcohol
   d. Phenol
   e. None of the above

Short Answer

5. Provide an unambiguous structural formula for (Z)-2-octen-4-ol.

6. Provide a systematic IUPAC name for the following compound.

Topic Test 1: Answers

1. **False.** Cyclohexanol is a secondary alcohol.

2. **True.** The three compounds mentioned are similar in size and shape differing only in whether the substituent at carbon 2 is CH_3, Cl, or OH. In the case of alcohol, hydrogen bonding between water molecules and the alcohol enhances the solubility.

3. **d.** Glycerol (1,2,3-propanetriol). Boiling point reflects the amount of energy required to overcome intermolecular attractions. Of all the possibilities listed, glycerol is the largest and has the greatest number of polar OH groups with which the molecules can hydrogen bond to one another.

4. **c.** Benzyl alcohol ($PHCH_2OH$). Responses a and b are both secondary alcohols, and d (phenol, PhOH) is technically not an alcohol at all.

5. 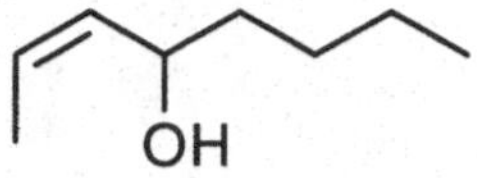 OH

6. *trans*-3-Phenylcyclobutanol

TOPIC 2: SOME PREPARATIONS OF ALCOHOLS

KEY POINTS

✓ *How are alcohols made from alkenes?*

✓ *What reducing agents will convert a carbonyl compound to an alcohol?*

✓ *What alcohol results from reduction of a given aldehyde or ketone?*

✓ *What alcohol results from reduction of a given ester or carboxylic acid?*

Alkenes can be hydrated with acid catalyst or through **oxymercuration** to give a Markovnikov alcohol. The **hydroboration** strategy provides a route to non-Markovnikov alcohols. These transformations are illustrated below for a generic monosubstituted alkene, but more highly substituted and cyclic alkenes react similarly.

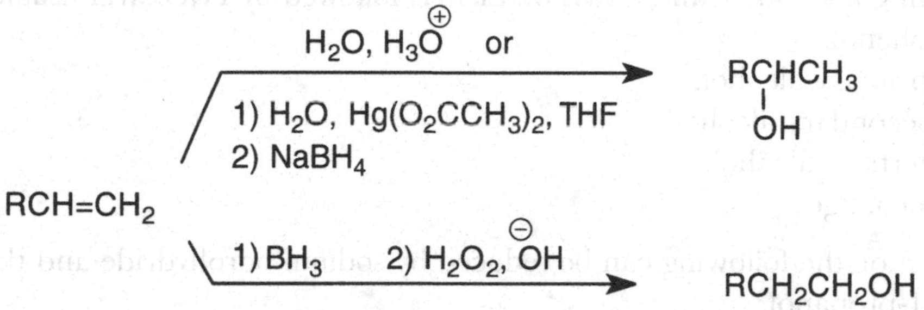

$$H_2O, H_3\overset{\oplus}{O} \quad or$$

$$1) \ H_2O, \ Hg(O_2CCH_3)_2, \ THF$$
$$2) \ NaBH_4$$

$$RCH{=}CH_2$$

$$1) \ BH_3 \quad 2) \ H_2O_2, \ \overset{\ominus}{O}H$$

$$RCHCH_3$$
$$\overset{|}{O}H$$

$$RCH_2CH_2OH$$

Alcohols can be synthesized from aldehydes and ketones by reduction with **sodium borohydride** or **lithium aluminum hydride** followed by aqueous acid. The mechanism for this transformation is discussed in Chapter 5, but for now note that these reagents add two hydrogens across the carbonyl pi bond. Aldehydes yield primary alcohols and ketones yield secondary alcohols.

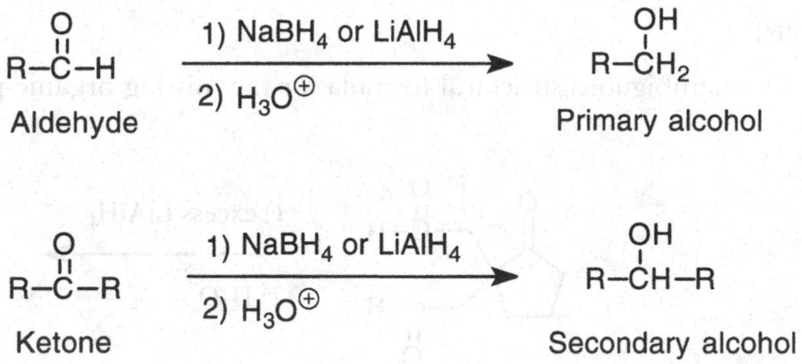

$$\begin{array}{c} O \\ \parallel \\ R{-}C{-}H \end{array} \xrightarrow[\text{2) } H_3O^{\oplus}]{\text{1) } NaBH_4 \text{ or } LiAlH_4} \begin{array}{c} OH \\ \mid \\ R{-}CH_2 \end{array}$$

Aldehyde   Primary alcohol

$$\begin{array}{c} O \\ \parallel \\ R{-}C{-}R \end{array} \xrightarrow[\text{2) } H_3O^{\oplus}]{\text{1) } NaBH_4 \text{ or } LiAlH_4} \begin{array}{c} OH \\ \mid \\ R{-}CH{-}R \end{array}$$

Ketone   Secondary alcohol

Sodium borohydride is not strong enough to reduce carboxylic acids and esters, but these can be converted to primary alcohols with lithium aluminum hydride followed by aqueous acid. Note that the carbonyl carbon becomes the CH_2 group that bears the OH in the final product.

$$\underset{\text{R-C-OH}}{\overset{\text{O}}{\|}} \xrightarrow{\text{LiAlH}_4} \xrightarrow{\text{H}_3\overset{\oplus}{\text{O}}} \text{R-CH}_2\text{OH}$$

$$\underset{\text{R-C-OR}'}{\overset{\text{O}}{\|}} \xrightarrow{\text{LiAlH}_4} \xrightarrow{\text{H}_3\overset{\oplus}{\text{O}}} \text{R-CH}_2\text{OH} \quad (\text{+ HOR}')$$

Topic Test 2: Some Preparations of Alcohols

True/False

1. Sodium borohydride followed by aqueous acid will reduce a ketone to an aldehyde.

2. Reduction of a carboxylic acid with sodium borohydride followed by aqueous acid will yield a tertiary alcohol.

Multiple Choice

3. Treating a ketone with $NaBH_4$ or $LiAlH_4$ followed by H_3O^+ will usually produce
   a. a phenol.
   b. a primary alcohol.
   c. a secondary alcohol.
   d. a tertiary alcohol.
   e. no change.

4. Which of the following can be reduced by sodium borohydride and then aqueous acid to give 1-propanol?
   a. $CH_3CH{=}CH_2$
   b. $CH_3CH_2CO_2H$
   c. $CH_3CH_2CH{=}O$
   d. All of the above
   e. None of the above

Short Answer

5. Provide an unambiguous structural formula for the missing organic product.

6. Show the reagents and/or conditions one could use to carry out the transformation indicated. More than one step may be required.

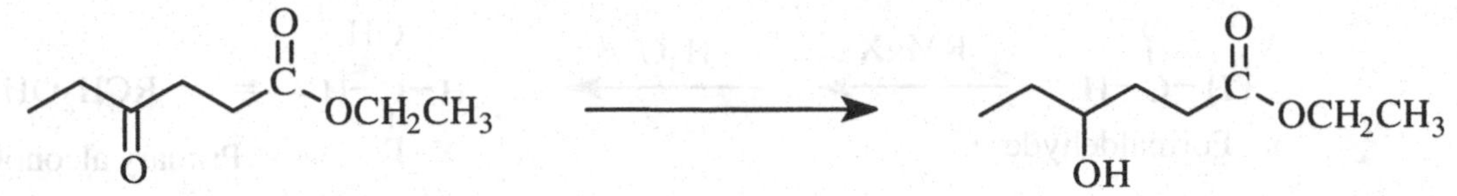

$$\text{O}$$
...C...OCH$_2$CH$_3$ → ...C...OCH$_2$CH$_3$ (OH)

Topic Test 2: Answers

1. **False.** Ketones are reduced to secondary alcohols under these conditions.

2. **False.** Sodium borohydride is too weak to reduce a carboxylic acid, and besides, reduction of a carboxylic acid would lead to a primary alcohol (not tertiary).

3. **c.** a secondary alcohol

4. **c.** $CH_3CH_2CH{=}O$. Sodium borohydride is too weak to reduce the carboxylic acid in response b or the alkene in response a. Even if the alkene in answer a reacted with the aqueous acid in the second step, the resulting product would be 2-propanol (Markovnikov) rather than 1-propanol.

5. 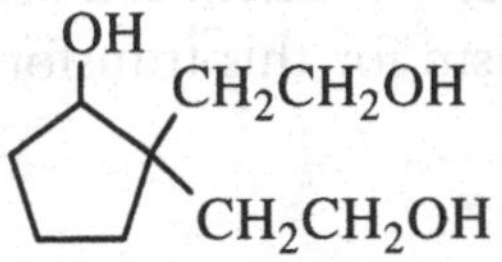 (structure: cyclopentane ring with OH and CH$_2$CH$_2$OH and CH$_2$CH$_2$OH substituents)

Note that the ketone becomes a secondary alcohol and the two aldehydes become primary alcohols under these conditions.

6. Use NaBH$_4$ followed by H_3O^+. Do not use LiAlH$_4$ because that reagent is strong enough to reduce the ester also.

TOPIC 3: ALCOHOLS FROM GRIGNARD REAGENTS

KEY POINTS

✓ *What is the product of a Grignard reagent reacting with formaldehyde?*

✓ *What is the product of a Grignard reagent reacting with any other aldehyde?*

✓ *What is the product of a Grignard reagent reacting with a ketone?*

✓ *What is the product of a Grignard reagent reacting with an ester?*

✓ *What is the product of a Grignard reagent reacting with epoxide?*

A variety of alcohols can be produced from **Grignard reagents** and appropriate electrophiles. When **formaldehyde** ($H_2C{=}O$) is reacted with a Grignard reagent followed by aqueous acid, the Grignard carbon ends up attached to the former carbonyl and the oxygen becomes an OH. Other aldehydes ($RCH{=}O$) and ketones ($R_2C{=}O$) undergo analogous transformations. In general, primary alcohols can be obtained from formaldehyde and a Grignard reagent. Other aldehydes and ketones lead to secondary and tertiary alcohols, respectively, as shown below.

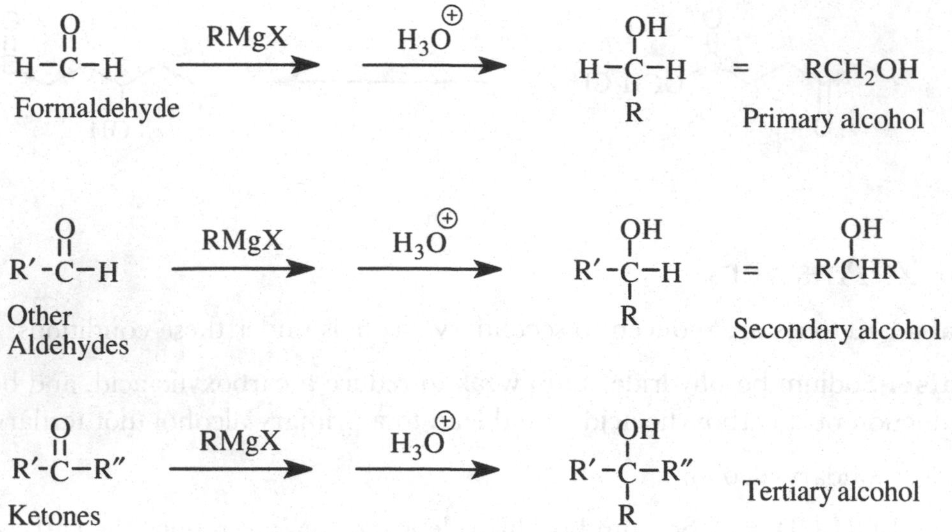

$$\underset{\text{Formaldehyde}}{H-\overset{\overset{\displaystyle O}{\|}}{C}-H} \xrightarrow{\text{RMgX}} \xrightarrow{H_3O^{\oplus}} \underset{\text{Primary alcohol}}{H-\overset{\overset{\displaystyle OH}{|}}{\underset{\underset{\displaystyle R}{|}}{C}}-H} = RCH_2OH$$

$$\underset{\substack{\text{Other} \\ \text{Aldehydes}}}{R'-\overset{\overset{\displaystyle O}{\|}}{C}-H} \xrightarrow{\text{RMgX}} \xrightarrow{H_3O^{\oplus}} R'-\overset{\overset{\displaystyle OH}{|}}{\underset{\underset{\displaystyle R}{|}}{C}}-H = \underset{\text{Secondary alcohol}}{R'CHR} \overset{OH}{}$$

$$\underset{\text{Ketones}}{R'-\overset{\overset{\displaystyle O}{\|}}{C}-R''} \xrightarrow{\text{RMgX}} \xrightarrow{H_3O^{\oplus}} R'-\overset{\overset{\displaystyle OH}{|}}{\underset{\underset{\displaystyle R}{|}}{C}}-R'' \quad \text{Tertiary alcohol}$$

When an ester is treated with excess Grignard reagent followed by aqueous acid, the carbonyl carbon becomes a tertiary alcohol carbon on which two identical groups (the R groups from RMgX) are attached. Frequently, the alkoxy (OR″) portion of the ester is discarded after reaction, but for completeness its fate is also shown below. The mechanism for this transformation is reviewed in Chapters 5 and 7.

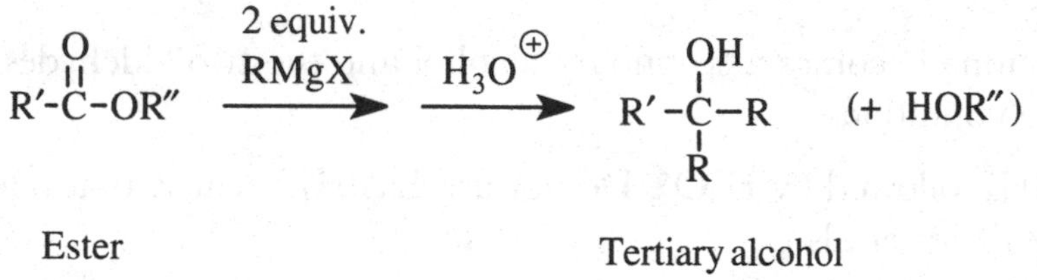

$$\underset{\text{Ester}}{R'-\overset{\overset{\displaystyle O}{\|}}{C}-OR''} \xrightarrow[\text{RMgX}]{\text{2 equiv.}} \xrightarrow{H_3O^{\oplus}} \underset{\text{Tertiary alcohol}}{R'-\overset{\overset{\displaystyle OH}{|}}{\underset{\underset{\displaystyle R}{|}}{C}}-R} \quad (+ \; HOR'')$$

Reaction of **epoxide** with a Grignard reagent followed by aqueous acid leads to a primary alcohol that is two methylene units longer than the R from RMgX. The mechanism is revisited in Chapter 4 and is similar to the S_N2 process.

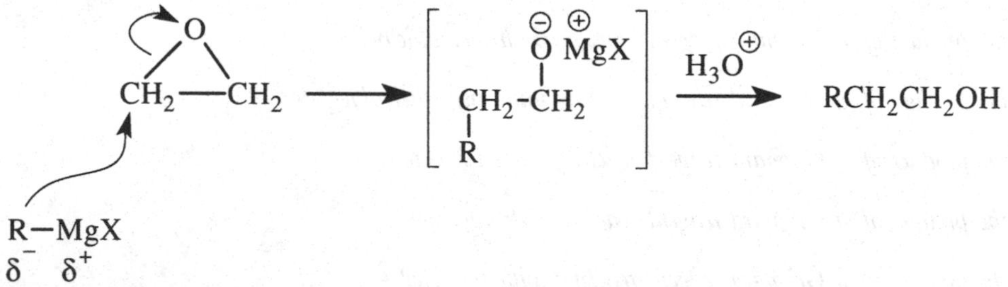

$$\underset{\underset{\displaystyle \delta^- \; \delta^+}{R-MgX}}{\overset{\displaystyle \overset{O}{\overbrace{}}}{CH_2-CH_2}} \longrightarrow \left[\underset{\underset{\displaystyle R}{|}}{\overset{\overset{\displaystyle \overset{\ominus}{O} \; \overset{\oplus}{MgX}}{|}}{CH_2-CH_2}} \right] \xrightarrow{H_3O^{\oplus}} RCH_2CH_2OH$$

Topic Test 3: Alcohols from Grignard Reagents

True/False

1. Phenylmagnesium bromide (PhMgBr) reacts with formaldehyde ($H_2C{=}O$) followed by acid to yield phenol.

2. Tertiary alcohols can be prepared from ketones and Grignard reagents.

Multiple Choice

3. Which of the following could have been prepared from a Grignard reagent and an ester?
   a. 1-Butanol
   b. 2-Butanol
   c. *tert*-Butyl alcohol
   d. All of the above
   e. None of the above

4. Which of the following could have been prepared from a Grignard reagent and epoxide?
   a. 1-Butanol
   b. 2-Butanol
   c. *tert*-Butyl alcohol
   d. All of the above
   e. None of the above

Short Answer

Provide unambiguous structural formulas for the missing organic compounds.

5.

6.

Topic Test 3: Answers

1. **False.** The product of that reaction sequence would be the primary alcohol phenyl-methanol (benzyl alcohol, $PhCH_2OH$).

2. **True.** The tertiary alcohol will bear three alkyl groups around the central COH. Two of those groups were what formerly surrounded the carbonyl in the starting ketone and the third group comes from RMgX.

3. **c.** *tert*-Butyl alcohol. Recall that excess RMgX reacts with an ester to give a tertiary alcohol bearing two identical alkyl groups around the COH. In this case, excess methyl Grignard (CH_3MgX) could have reacted with any CH_3CO_2R.

4. **a.** 1-Butanol. Recall that RMgX reacts with epoxide to yield, after H_3O^+ treatment, a primary alcohol of the form RCH_2CH_2OH. In this case ethyl Grignard and epoxide would lead to 1-butanol.

5.
$$CH_3\text{–}CH\text{–}\overset{\overset{\displaystyle OH}{|}}{CH}\text{–}\bigcirc\text{–}CH_3$$
with CH_3 groups

6. cyclohexanone structure

TOPIC 4: SOME REACTIONS OF ALCOHOLS

KEY POINTS

✓ *What reagents are used to dehydrate alcohols to alkenes?*

✓ *How are alcohols converted to alkyl halides?*

✓ *What reagents are used to oxidize alcohols?*

✓ *What product results from oxidation of a given primary alcohol?*

✓ *What product results from the oxidation of a given secondary alcohol?*

Alcohols undergo acid-catalyzed **dehydration** to yield alkenes when treated with concentrated sulfuric or phosphoric acid. The mechanism is E1 and is the reverse of the acid-catalyzed hydration of alkenes. The regiochemistry of the major product is predictable using Zaitsev's rule. Compounds for which strong acid and heat or carbocation rearrangements are a problem can be dehydrated with phosphorous oxychloride via an E2 mechanism.

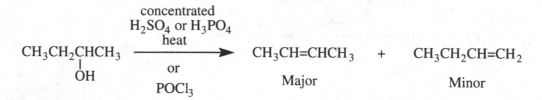

$$CH_3CH_2\underset{\underset{\displaystyle OH}{|}}{CH}CH_3 \xrightarrow[\substack{\text{or} \\ POCl_3}]{\substack{\text{concentrated} \\ H_2SO_4 \text{ or } H_3PO_4 \\ \text{heat}}} CH_3CH{=}CHCH_3 \quad + \quad CH_3CH_2CH{=}CH_2$$

Major   Minor

Tertiary alcohols can be converted to the corresponding alkyl chlorides or bromides with HCl and HBr, respectively. The mechanism is S_N1. Methyl, primary, and secondary alcohols are converted to RCl or RBr via S_N2 mechanisms with **thionyl chloride** or **phosphorous tribromide**, respectively.

$$\overset{|}{\underset{|}{C}}\text{–}OH$$

$$\xrightarrow{HX} \quad \overset{|}{\underset{|}{C}}\text{–}X \quad \substack{(X = Cl \text{ or } Br, \text{ best} \\ \text{for tertiary ROH, } S_N1)}$$

$$\xrightarrow{SOCl_2} \quad \overset{|}{\underset{|}{C}}\text{–}Cl$$

$$\xrightarrow[\text{ether}]{PBr_3} \quad \overset{|}{\underset{|}{C}}\text{–}Br$$

Oxidation of alcohols can be carried out with a variety of reagents, including **Jones' reagent** (CrO_3, H_2SO_4, H_2O), $Na_2Cr_2O_7$, $KMnO_4$, and **pyridinium chlorochromate** (PCC). Primary alcohols are oxidized to carboxylic acids with most oxidizing agents, but when PCC is used cold, the oxidation can be stopped at the aldehyde. Secondary alcohols oxidize to ketones with any of these reagents, but tertiary alcohols do not oxidize under these conditions.

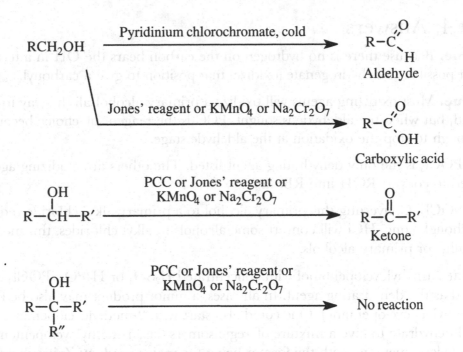

RCH$_2$OH → (Pyridinium chlorochromate, cold) → R–C(=O)H **Aldehyde**

RCH$_2$OH → (Jones' reagent or KMnO$_4$ or Na$_2$Cr$_2$O$_7$) → R–C(=O)OH **Carboxylic acid**

R–CH(OH)–R' → (PCC or Jones' reagent or KMnO$_4$ or Na$_2$Cr$_2$O$_7$) → R–C(=O)–R' **Ketone**

R–C(OH)(R'')–R' → (PCC or Jones' reagent or KMnO$_4$ or Na$_2$Cr$_2$O$_7$) → **No reaction**

Topic Test 4: Some Reactions of Alcohols

True/False

1. Tertiary alcohols do not generally react with PCC or Jones' reagent.

2. Gentle oxidation of a primary alcohol with PCC yields an aldehyde, but more vigorous oxidation with Jones' reagent leads to a carboxylic acid.

Multiple Choice

3. Which reagent below can be used to convert cyclohexanol into cyclohexene?
   a. PBr$_3$
   b. PCC
   c. POCl$_3$
   d. KMnO$_4$
   e. None of the above

4. Which reagent below would be best for converting 3-phenyl-1-propanol into 3-phenyl-1-chloropropane?
   a. conc. HCl
   b. POCl$_3$
   c. SOCl$_2$
   d. Cl$_2$
   e. CCl$_4$

Short Answer

Show how one could prepare each of the following from an alcohol.

5. 1-Methylcyclopentene

6. 2-Chloro-2-methyl-2-butanol

Topic Test 4: Answers

1. **True.** Because there is no hydrogen on the carbon bears the OH in a tertiary alcohol, it is not possible to dehydrogenate (oxidize) that position to give a carbonyl.

2. **True.** Most oxidizing agents will oxidize a primary alcohol all the way to a carboxylic acid, but when the aldehyde is sought, PCC is the reagent of choice because it is gentle enough to stop the oxidation at the aldehyde stage.

3. **c.** $POCl_3$ is the only dehydrating agent listed. The others are oxidizing agents (b and d) or used to convert ROH into RBr (a).

4. **c.** $SOCl_2$. Converting this primary alcohol to a primary alkyl chloride requires an S_N2. Although conc. HCl will convert some alcohols to alkyl chlorides, this method gives slow results for primary alcohols.

5. Heat 1-methylcyclopentanol with concentrated H_2SO_4 or H_3PO_4. $POCl_3$ could also be used as the dehydrating agent. In all cases, a minor product may also be formed (methylenecyclopentane). One could also start with 2-methylcyclopentanol, which will dehydrate to give a mixture of regioisomers (i.e., 1-methylcyclopentene and 3-methylcyclopentene with the former being the major product: Zaitsev's rule).

6. This tertiary alkyl halide could most easily be prepared by treating 2-methyl-2-butanol with concentrated hydrochloric acid. Recall that this method works best for tertiary alcohols because a carbocation intermediate is involved. Thionyl chloride would not be a good choice for this conversion because the mechanism in that case is S_N2 that is unfavorable for tertiary substrates.

TOPIC 5: PREPARATIONS AND REACTIONS OF PHENOLS

KEY POINTS

✓ *How can phenols be prepared through nucleophilic aromatic substitution?*

✓ *What is the cumene hydroperoxide synthesis of phenols?*

✓ *How can a phenol be made via an aryldiazonium?*

✓ *What is the general reactivity of phenols in EAS?*

✓ *How acidic are phenols compared with alcohols or carboxylic acids, and so forth?*

If an aromatic compound bears a suitable leaving group, it can undergo a **nucleophilic aromatic substitution** by hydroxide nucleophile to yield a phenol. These reactions occur by addition–elimination mechanisms if the aromatic ring bears electron-withdrawing substituents. Without electron withdrawing groups, an elimination–addition mechanism (via benzyne) occurs under vigorous conditions (Chapter 1).

$$Ar-L \xrightarrow{\overset{\ominus}{O}H} Ar-OH \quad (+ \; L^{\ominus})$$

An industrial preparation of phenol begins with cumene (isopropylbenzene) and is known as the **cumene hydroperoxide synthesis**. First, cumene is heated in air whereupon the benzylic hydrogen is oxidized to give cumene hydroperoxide. This unstable intermediate is then treated with aqueous acid and decomposes to yield phenol and acetone.

We will see in Chapter 9 that aromatic amines react with nitrous acid to give **aryldiazonium ions** that in turn can be converted to phenols with hot aqueous acid.

$$Ar-NH_2 \xrightarrow{HONO} Ar-\overset{\oplus}{N_2} \xrightarrow{H_3O^{\oplus}} Ar-OH$$

The OH group of a phenol strongly activates the ring for *ortho/para*-directed EAS. For brominations, this effect is so strong that no catalyst is needed and all open *ortho* and *para* positions become brominated. This EAS activation can be attenuated by converting the OH of the phenol into an ester using a method we cover in detail in Chapter 7.

Phenols are more acidic than alcohols but less acidic than carboxylic acids or mineral acids. Resonance delocalization of the negative charge into the ring makes the conjugate base of a phenol more stable and explains the enhanced acidity.

$$ROH < ArOH < RCO_2H < HNO_3$$

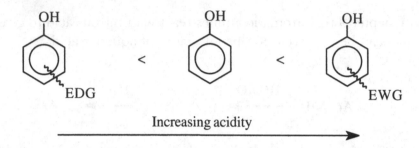

Electron-withdrawing groups on phenols increase their acidity, whereas electron-donor groups repress the acidity. The effect can be inductive or via resonance, with the latter usually being more important.

OH   OH   OH

EDG   <   <   EWG

Increasing acidity →

Topic Test 5: Preparation and Reactions of Phenols

True/False

1. Phenol is more reactive than toluene or cumene in EAS reactions.

2. 1-Chloro-2,4-dinitrobenzene can be converted to 2,4-dinitrophenol via nucleophilic aromatic substitution with hydroxide as the nucleophile.

Multiple Choice

3. Which of the following can be converted into phenol?
   a. Ph—Cl
   b. $PhCH(CH_3)_2$
   c. PhN_2^+
   d. All of the above
   e. None of the above

4. Which compound is most acidic?
   a. Phenol
   b. p-Nitrophenol
   c. p-Methoxyphenol
   d. p-Ethylphenol
   e. Cyclohexanol

Short Answer

What products would result from reaction of *ortho*-ethylphenol with each of the following reagents?

5. Excess Br_2, $FeBr_3$

6. KOH (aq), 25°C

Topic Test 5: Answers

1. **True.** The OH group of phenol is a strong electron donor (resonance) that enhances the electron density of the ring, making it more reactive toward electrophiles. Alkyl groups such as the methyl or isopropyl of toluene and cumene are weak electron donors and as such activate the aromatic ring for EAS only slightly.

2. **True.** Hydroxide attacks carbon number 1, making a negatively charged intermediate. The presence of the two nitro groups allows the negative charge on the intermediate to be resonance stabilized more effectively, and the reaction takes place at fairly mild temperatures. (Draw several resonance forms of the intermediate, including at least one in which the negative charge is not on a ring atom.)

3. **d.** All of the above. PhCl reacts with hydroxide at high temperature via benzene. $PhCH(CH_3)_2$ is the starting material for the cumene hydroperoxide synthesis. Phenyldiazonium reacts with aqueous acid to give phenol.

4. **b.** p-Nitrophenol is most acidic. Cyclohexanol is an alcohol and is less acidic. Phenols are most acidic when they bear electron withdrawing substituents such as nitro. Choices c and d bear OCH_3 and ethyl, respectively. Those are electron-donating substituents that render the phenol less acidic than phenol itself.

5. The major product would be 2,4-dibromo-6-ethylphenol, but one might also observe some small amounts of monobromo or tribromo products.

6. This acid-base reaction will lead to water and the potassium salt of phenol (i.e., potassium phenoxide, PhO^-K^+).

TOPIC 6: THIOLS

KEY POINTS

✓ *What is the structure of a thiol?*

✓ *What are some properties of thiols?*

✓ *How are thiols named?*

✓ *How and from what are thiols prepared?*

✓ *What is the relationship between a disulfide and a thiol?*

Thiols are structural analogues of alcohols except a sulfur replaces the oxygen. These are unpopular compounds mostly due to their characteristic unpleasant odor. The active components in a skunk's defense scent are thiols. Many utility companies spike their odorless natural gas with trace thiols that produce an aroma that alerts consumers to leaks. Because sulfur is less electronegative than oxygen, the polarity of thiols is somewhat less than alcohols of similar size and shape. The systematic IUPAC names of thiols are obtained by adding the suffix "thiol" to the parent alkane name. The common names of thiols are based on the attraction of the sulfur for

Table 3.2 Names and Structures of Some SH Compounds

STRUCTURE	COMMON	IUPAC
CH$_3$SH	Methyl mercaptan	Methanethiol
CH$_3$CH$_2$SH	Ethyl mercaptan	Ethanethiol
(CH$_3$)$_2$CHSH	Isopropyl mercaptan	2-Propanethiol
(cyclohexyl)—SH	Cyclohexyl mercaptan	Cyclohexanethiol
HS—(benzaldehyde structure)	m-Mercaptobenzaldehyde	

mercury. The first word is the name of the alkyl group followed by the word "**mercaptan**" (mercury capture). Some compounds are named such that the SH group is a mercapto substituent on the parent structure. **Table 3.2** shows examples of RSH compound structures and names.

Thiols can be synthesized from alkyl halides with hydrosulfide anion as a nucleophile. Because sulfur is more nucleophilic and less basic than oxygen, elimination is less of a competing side reaction; however, better yields are obtained if thiourea followed by aqueous base is used.

$$R{-}X \quad + \quad \overset{\ominus}{S}H \quad \longrightarrow \quad R{-}SH \quad + \quad X^{\ominus}$$

$$\uparrow \;\; H_2O, OH^{\ominus}$$

$$R{-}X \quad + \quad \underset{\underset{NH_2}{H_2N}}{\overset{S}{\underset{\|}{C}}} \quad \longrightarrow \quad \left[R{-}\overset{\oplus}{S}{=}C(NH_2)_2 \right] X^{\ominus}$$

Mild oxidizing agents such as Br$_2$ or I$_2$ can be used to convert thiols to **disulfides**, RSSR. The reaction is reversible with zinc metal and acid; thus, thiols and disulfides can be interconverted through oxidation and reduction reactions.

$$2\,RSH \quad \underset{Zn,\,H_3O^{\oplus}}{\overset{Br_2\ or\ I_2}{\rightleftharpoons}} \quad R\text{-}S\text{-}S\text{-}R$$

Topic Test 6: Thiols

True/False

1. The IUPAC name for *t*-butyl mercaptan is 2-methyl-2-propanethiol.

2. Thiols have the general structure R-S-S-R.

Multiple Choice

3. Which reagent(s) below will convert thiols into disulfides?
   a. Thiourea followed by aqueous alkali.
   b. Zinc metal in aqueous acid.
   c. Molecular bromine, Br_2.
   d. All of the above.
   e. None of the above.

4. Which of the following applies to thiols?
   a. They can be made from alkyl halides.
   b. They often stink.
   c. They can be interconverted with disulfides.
   d. All of the above.
   e. None of the above.

Short Answer

5. Show the reagents and/or conditions one could use to carry out the transformation below. More than one step may be required.

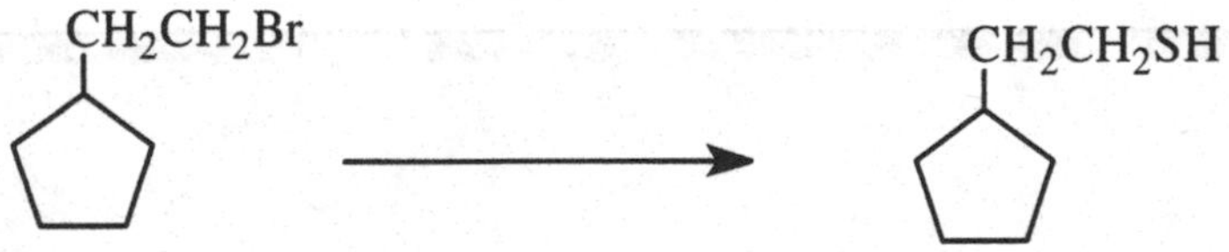

6. Complete the following reaction scheme with an unambiguous structural formula.

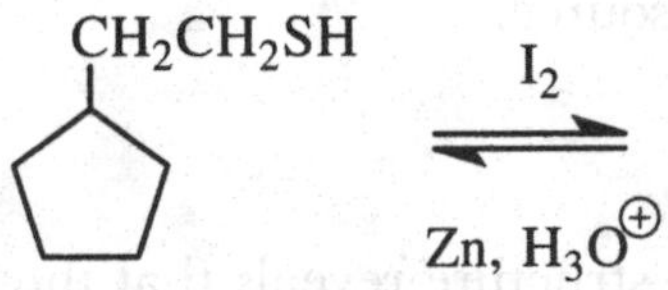

Topic Test 6: Answers

1. **True.** The longest chain bearing the SH group is 3 carbons. The SH group and a methyl group are attached to the second carbon.

2. **False.** Thiols have the general structure R-S-H. The structural formula shown in the problem is a disulfide.

3. **c.** Molecular bromine, Br_2. Reagent a is for converting RX into RSH and reagent b is for reducing disulfides to thiols (the reverse of the process in the problem statement).

4. **d.** All of the above

5. An S_N2 reaction with HS^- will work, but better yields can be obtained using $(NH_2)_2C{=}S$ (thiourea) followed by aqueous alkali (HO^-, H_2O).

6. 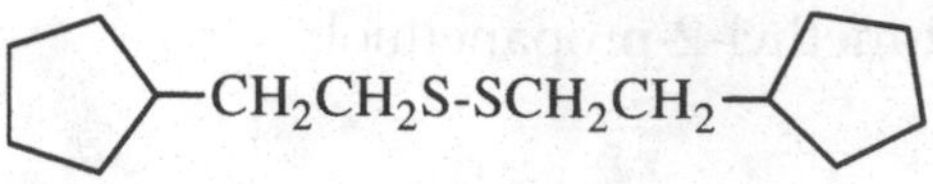 ⬠–CH_2CH_2S-SCH_2CH_2–⬠

APPLICATION

The active ingredient in alcoholic beverages such as beer and wine is ethanol. Most of it comes from fermentation wherein yeast breaks down sugars to yield ethanol and carbon dioxide.

$$C_6H_{12}O_6 \xrightarrow{\text{yeast}} CH_3CH_2OH + CO_2$$
a sugar   ethanol

Like all primary alcohols, ethanol is easily oxidized to a carboxylic acid in the presence of suitable oxidizing agents. One popular oxidizer is dichromate ion, $Cr_2O_7^{2-}$. This bright orange reagent changes color to drab green when it oxidizes ethanol to acetic acid. This visual change provides the chemical basis for the roadside test to measure blood alcohol known as the "breathalyzer."

$$3CH_3CH_2OH + 2Cr_2O_7^{2-} + 16H_3O^+ \rightarrow 3CH_3CO_2H + 4Cr^{3+} + 27H_2O$$
ethanol   dichromate   acetic acid   chromium (III)
            (orange)   (green)

A more modern technology for easy measurement of blood alcohol is based on infrared spectroscopy (Chapter 2).

DEMONSTRATION PROBLEM

Show how one could synthesize 4-ethyl-2-methyl-4-heptanol using alcohols with five or fewer carbons as the only carbon source.

Solution

Translating the name into a structure reveals that the target molecule is a tertiary alcohol that bears three different alkyl groups around the alcohol carbon. As usual, the most reliable way to reason through a multistep synthesis problem of this type is to work through it backward. The tertiary alcohol was likely made from reaction of a Grignard reagent with a ketone. The ketone could have resulted from oxidation of a secondary alcohol. That secondary alcohol can be made from a Grignard reagent and an aldehyde. The aldehyde can be made by gentle oxidation of a primary alcohol. Under the stipulation that alcohols are the only carbon source, we must also prepare the Grignard reagents from alcohols separately. There is more than one way to prepare this compound using familiar reactions. One method is shown here.

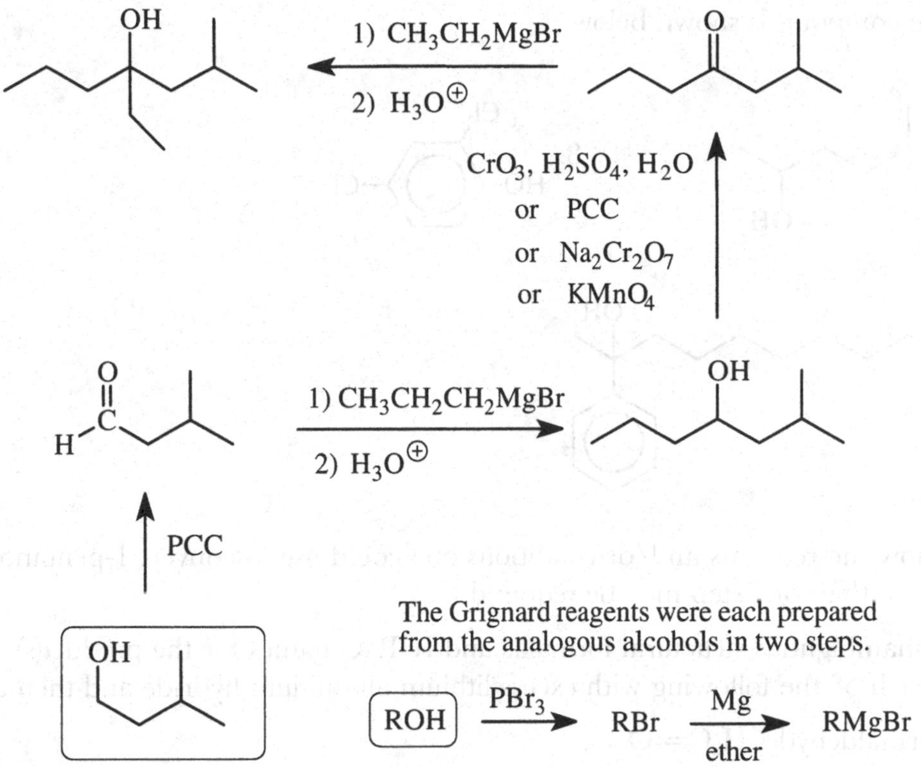

1) CH_3CH_2MgBr
2) $H_3O^{\oplus}$

CrO_3, H_2SO_4, H_2O
or   PCC
or $Na_2Cr_2O_7$
or $KMnO_4$

1) $CH_3CH_2CH_2MgBr$
2) $H_3O^{\oplus}$

PCC

The Grignard reagents were each prepared
from the analogous alcohols in two steps.

$$\boxed{ROH} \xrightarrow{PBr_3} RBr \xrightarrow[\text{ether}]{Mg} RMgBr$$

Chapter Test

1. Which of the following applies to secondary alcohols?
   a. They are more acidic than phenols.
   b. They react with $NaBH_4$ and then aqueous acid to yield aldehydes or ketones.
   c. Their boiling points are higher than those of similar alkanes or alkyl halides.
   d. All of the above
   e. None of the above

2. Reaction of isopropyl mercaptan with I_2 will most likely yield
   a. isopropyl alcohol
   b. 2-iodopropane
   c. a ketone
   d. $(CH_3)_2CHSSCH(CH_3)_2$
   e. None of the above

3. Which of the following reactions will likely yield a phenol?
   a. An aryldiazonium (ArN_2^+) treated with aqueous acid
   b. Nucleophilic aromatic substitution by hydroxide on ArCl
   c. Heating $ArCH(CH_3)_2$ in air and then aqueous acid
   d. All of the above
   e. None of the above

Provide unambiguous structural formulas for each of the following.

4. *trans*-1,2-Cyclohexanediol

5. Allyl mercaptan

6. *m*-cyclopropylphenol

Name the compounds shown below

7.

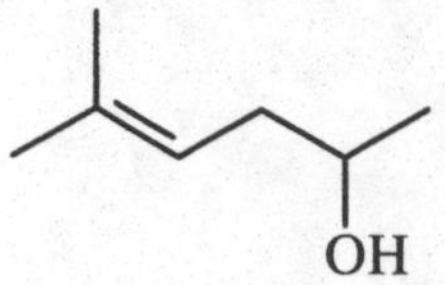

8.

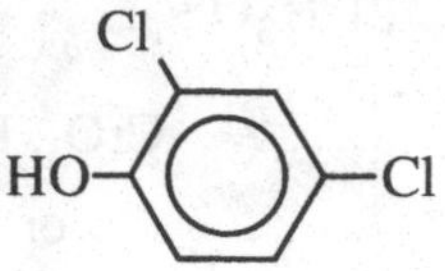

9.

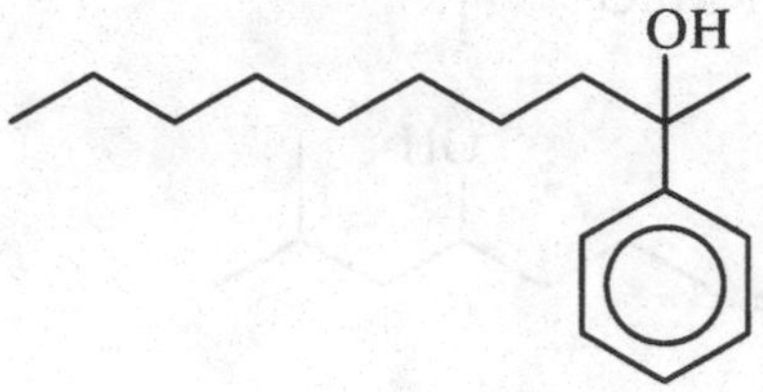

10. Show the reagents and/or conditions one could use to convert 1-pentanol into 2-pentanol. More than one step may be required.

Provide unambiguous structural formulas and IUPAC names for the product(s) obtained from treating each of the following with excess lithium aluminium hydride and then aqueous acid.

11. Formaldehyde, $H_2C\!=\!O$

12. Octanal, $CH_3(CH_2)_6CH\!=\!O$

13. 4-Heptanone, $(CH_3CH_2CH_2)_2C\!=\!O$

14. The ester ethyl acetate, $CH_3CO_2CH_2CH_3$

Provide unambiguous structural formulas and IUPAC names for the organic product(s) obtained from treating each of the following with excess ethyl Grignard (ethylmagnesium halide, CH_3CH_2MgX) and then aqueous acid.

15. Formaldehyde, $H_2C\!=\!O$

16. Octanal, $CH_3(CH_2)_6CH\!=\!O$

17. 4-Heptanone, $(CH_3CH_2CH_2)_2C\!=\!O$

18. The ester ethyl acetate, $CH_3CO_2CH_2CH_3$

19. Epoxide, the three-membered ring cyclic ether, C_2H_4O

20. Show how one could synthesize 1-cyclohexylcyclohexanol using cyclohexanol as the only carbon source.

21. Explain with words and pictures why ethanol is more water soluble than ethane or chloroethane.

22. What reagents and/or conditions can be used to convert a primary alcohol into an aldehyde?

23. List at least two sets of reagents and/or conditions that will convert a primary alcohol into a carboxylic acid.

24. Phenol dissolves in 5% aqueous NaOH yet cyclohexanol is not readily soluble in dilute alkali. Explain this observation and write a reaction and/or show structural formulas as needed.

25. The two compounds below undergo EAS at different rates. Predict which one is faster and explain. Draw resonance forms to support your explanation.

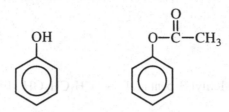

26. Write a reaction scheme that shows how $CH_3CH_2SSCH_2CH_3$ can be converted to ethanethiol.

27. Show how one could make 1-butanethiol from 1-butanol. More than one step may be required.

28. Show how to make cyclopentylmethanol from an alkene.

29. Provide structural formulas for the organic products that result from heating 1-tert butyl-4-isopropylbenzene in air followed by aqueous acid.

30. Draw all reasonable products one would expect from warming 1,2-dimethylcyclopentanol with concentrated sulfuric acid. Indicate which is the major product.

Chapter Test: Answers

1. **c** 2. **d** 3. **d**

4. 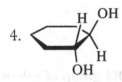   5. $HSCH_2CH=CH_2$ 6.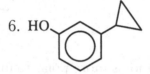

7. 5-Methyl-4-hexene-2-ol

8. 2,4-Dichlorophenol

9. 2-Phenyl-2-decanol

10.

~~~~~OH $\xrightarrow{POCl_3}$ ~~~~~ $\xrightarrow[\text{2) NaBH}_4]{\substack{\text{1) Hg(O}_2\text{CCH}_3)_2 \\ \text{H}_2\text{O, THF}}}$ ~~~~~OH

(Other reagents are possible but less reliable.)

11. CH_3OH, methanol

12. $CH_3(CH_2)_6CH_2OH$, 1-octanol

13. $(CH_3CH_2CH_2)_2CHOH$, 4-heptanol

14. CH_3CH_2OH, ethanol

15. $CH_3CH_2CH_2OH$, 1-propanol

16. $CH_3(CH_2)_6\overset{\overset{\displaystyle OH}{|}}{C}HCH_2CH_3$ 3-Decanol

17.

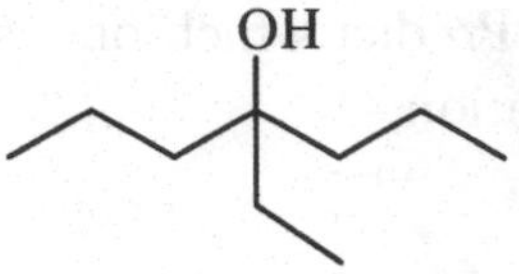

OH

4-Ethyl-4-heptanol

18.

OH

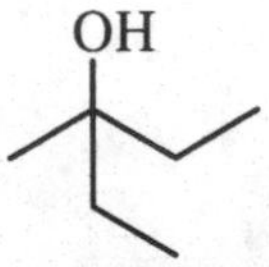

3-Methyl-3-pentanol + CH_3CH_2OH Ethanol

19. $CH_3CH_2CH_2CH_2OH$, 1-butanol

20.

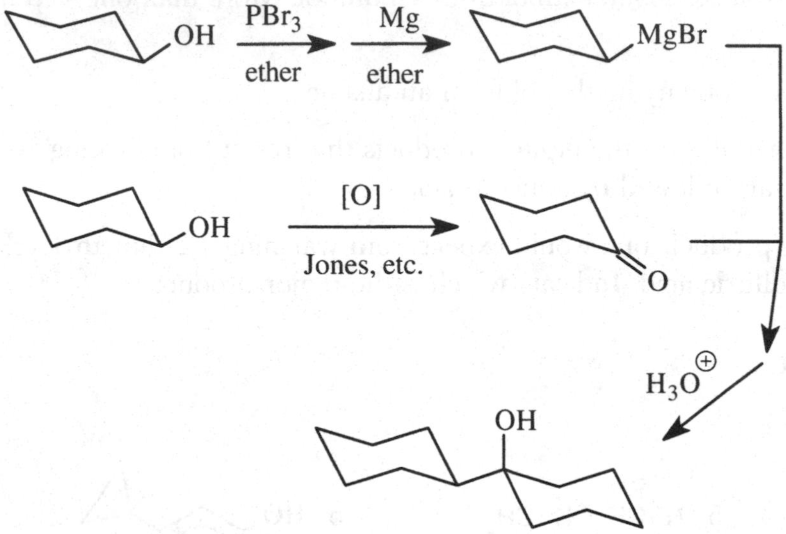

21. Ethanol is more polar than either ethane or chloroethane. The OH group of ethanol allows hydrogen bond donating or accepting as shown below.

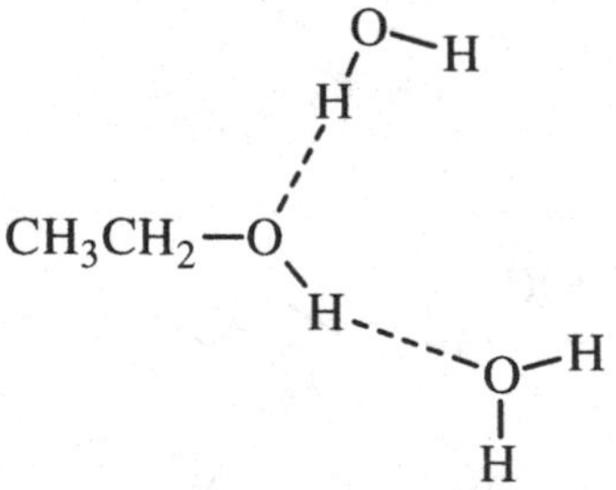

22. PCC

23. CrO_3, H_2SO_4, H_2O (Jones' reagent) or $K_2Cr_2O_7$, H_3O^+, or $KMnO_4$, others possible.

24. The conjugate base of phenol is more stable than that of cyclohexanol. In dilute aqueous alkali, phenol is acidic enough to react, whereas cyclohexanol is not.

$$PhOH + NaOH(aq) \rightarrow PhO^-Na^+(aq) + H_2O$$

25. Phenol is more reactive than the phenyl ester shown. For either compound, the cationic intermediate that results from *ortho* or *para* attachment of the electrophile is resonance stabilized by the lone pairs of electrons on oxygen. For the ester, the electron pairs on oxygen are also conjugated with carbonyl, which withdraws some of the electron density. Phenol is more reactive in EAS reactions.

26. $CH_3CH_2SSCH_2CH_3 + Zn, H_3O^+ \rightarrow 2CH_3CH_2SH$

27. $CH_3CH_2CH_2CH_2OH$ $\xrightarrow[\text{ether}]{PBr_3}$ $CH_3(CH_2)_3Br$ $\xrightarrow[\text{2) } H_2O, OH^{\ominus}]{\text{1) } S=C(NH_2)_2}$ $CH_3(CH_2)_3SH$

28.

29. $(CH_3)_3C-$$-OH$ + $CH_3-\overset{\overset{\textstyle O}{\|}}{C}-CH_3$

30.

Check Your Performance

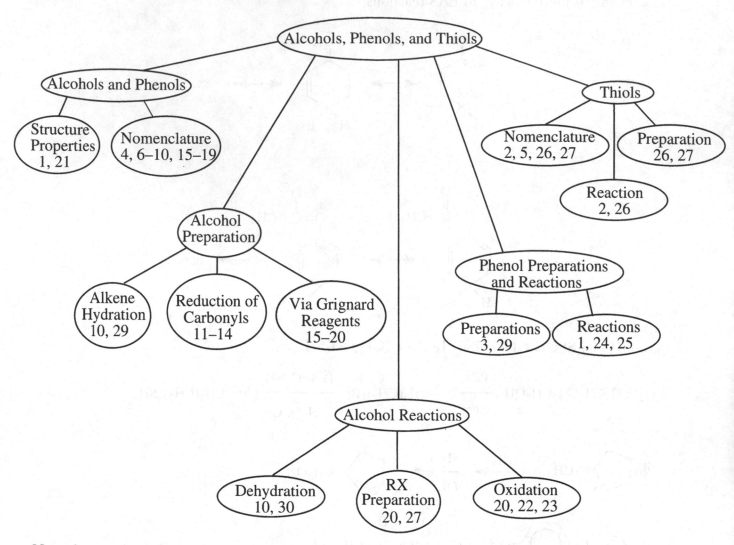

Note the number of questions in each grouping that you got wrong on the chapter test. Identify areas where you need further review and go back to relevant parts of this chapter.

CHAPTER 4

Ethers, Epoxides, and Sulfides

When most people say "ether" they mean one of the most common and simple members of this class, diethyl ether. At one time diethyl ether was widely used as an inhaled general anesthetic. It has a low boiling point, high vapor pressure at room temperature, and is highly flammable, making it poorly suited for use in a modern operating room containing many electrical appliances. In this chapter we survey the chemistry of ethers and related compounds.

ESSENTIAL BACKGROUND

- Alkane and alkyl group nomenclature
- Bond dipoles, molecular dipoles, and hydrogen bonding
- Markovnikov's rule for addition reactions to alkenes
- Oxymercuration
- Halohydrin formation
- Nucleophilic substitution on alkyl halides
- Competition of elimination and substitution

TOPIC 1: STRUCTURE, PROPERTIES, AND NOMENCLATURE

KEY POINTS

✓ *What are the structures of ethers, epoxides, and sulfides?*

✓ *How are ethers and sulfides named?*

✓ *How do the properties of ethers compare with those of alkanes or alcohols?*

✓ *What is the polarity of a typical ether?*

✓ *How do the structure and polarity of an ether explain its properties?*

The general form of an **ether** is R—O—R′, where R can be alkyl, vinyl, or aromatic but not any atom or group attached at an atom other than carbon. Note that the R groups need not be the same (although that is often the case), and for cyclic ethers the two R groups are connected. Three-membered ring cyclic ethers are sufficiently different from other ethers that they are separately designated as **epoxides**. Sulfur analogues of ethers are called **sulfides** and have the general structure R—S—R. Here too the R groups can be the same or different and can include alkyl and aryl groups.

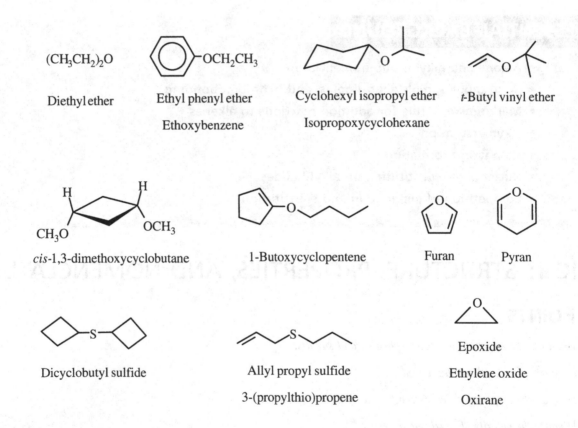

Ether   Epoxide   Sulfide

Ethers are named two ways. Simple ethers list the R groups (usually in alphabetical order) followed by the separate word "ether." Alternatively, ethers can be named as **alkoxy** (alkyl + oxygen) substituted parents. This method is especially useful for more complex molecules that contain the ether functional group. Some cyclic ethers have parent names from which other names are derived. Names for sulfides are similar to those of ethers except that the term "sulfide" is used in place of "ether," and an RS group is designated as an **alkylthio** rather than alkoxy. Parentheses are sometimes used to aid clarity. These nomenclatures are illustrated in the examples below.

$(CH_3CH_2)_2O$

Diethyl ether

$\text{—}OCH_2CH_3$

Ethyl phenyl ether

Ethoxybenzene

Cyclohexyl isopropyl ether

Isopropoxycyclohexane

t-Butyl vinyl ether

CH_3O OCH_3

cis-1,3-dimethoxycyclobutane

1-Butoxycyclopentene

Furan   Pyran

Dicyclobutyl sulfide

Allyl propyl sulfide

3-(propylthio)propene

Epoxide

Ethylene oxide

Oxirane

The high electronegativity of oxygen and its sp³ hybridization gives the ether linkage a permanent dipole around a bent geometry. There is no hydrogen bound directly to the oxygen, so ethers cannot be hydrogen bond donors but can be hydrogen bond acceptors. Ethers are generally less polar than alcohols but slightly more polar than alkanes and other hydrocarbons of similar size and shape. The boiling points and melting points of ethers are generally between those of similarly sized alcohols and hydrocarbons. Small ethers (those with four or fewer carbons) are at least partially water soluble.

Topic Test 1: Structure, Properties, and Nomenclature

True/False

1. The compound with the structure $CH_2=CH-O-CH=CH_2$ is divinyl ether.

2. Ethers are hydrogen bond acceptors and are therefore generally more soluble in water than hydrocarbons of similar size and shape.

Multiple Choice

3. The structure R—S—R', where R and R' are alkyl groups is a
   a. thiol.
   b. mercaptan.
   c. sulfide.
   d. disulfide.
   e. None of the above

4. Which of the following would have the lowest boiling point?
   a. $HOCH_2CH_2CH_2CH_2OH$
   b. $HOCH_2CH_2CH_2OCH_3$
   c. $HOCH_2CH_2OCH_2CH_3$
   d. $CH_3OCH_2CH_2OCH_3$
   c. These isomers all have nearly the same boiling point.

Short Answer

5. Provide an unambiguous structural formula for 2-methylthiofuran.

6. Name the following.

Topic Test 1: Answers

1. **True.** Ethers are named according to the groups around the oxygen followed by the second word "ether." The group $CH_2=CH-$ is called vinyl (or ethenyl).

2. **True.** The important intermolecular force leading to water solubility is hydrogen bonding between water and the ether ($H-O-H \cdots OR_2$)

3. **c**

4. **d.** $CH_3OCH_2CH_2OCH_3$. This diether is the only compound of the list that has no OH group(s) and therefore will not participate in hydrogen bonding.

5.

6. 3-Propoxycyclohexene (i.e., a cyclohexene bearing a propoxy (propyl + oxy) substituent at C3).

TOPIC 2: PREPARATION OF ETHERS, EPOXIDES, AND SULFIDES

KEY POINTS

✓ *How are symmetrical ethers prepared industrially?*

✓ *How could one prepare a given ether by alkoxymercuration?*

✓ *What reagents and reactions constitute the Williamson ether synthesis?*

✓ *How are epoxides synthesized?*

✓ *How are sulfides synthesized?*

Intermolecular dehydration of alcohols is used for the large-scale industrial synthesis of symmetrical alcohols. This method is not generally applicable to nonsymmetrical ethers because mixtures unavoidably result.

$$2R-O-H \xrightarrow[\text{heat}]{H_2SO_4} R-O-R + H_2O$$

$$ROH + HOR' \xrightarrow[\text{heat}]{H_2SO_4} ROR + ROR' + R'OR' \text{ (mixture)}$$

Markovnikov addition of water across an alkene can be carried out by oxymercuration. An analogous addition of alcohol called **alkoxymercuration** will give an ether. The mercuration reagent is usually mercury (II) trifluoroacetate in this case.

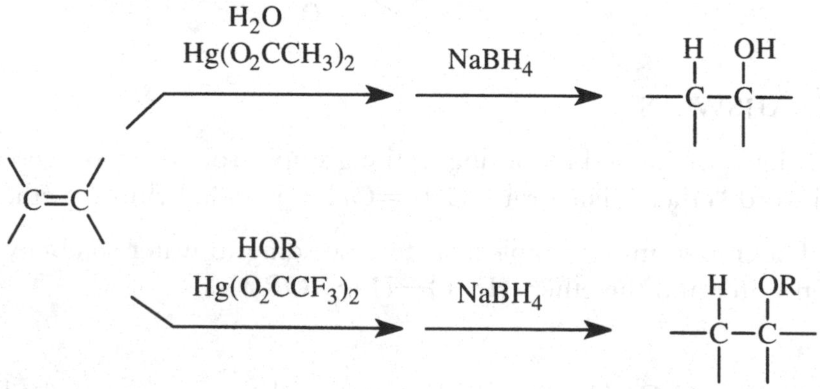

The most important laboratory synthesis of ethers involves an alkoxide nucleophile and an S_N2 attack on an alkyl halide. Both the halide and the alkoxide can be obtained from the corresponding alcohols as shown below. The S_N2 mechanism limits this method such that tertiary alkyl halides cannot be used and best results are realized with methyl or primary halides.

$$\text{ROH} \xrightarrow[\substack{\text{NaH or}\\ \text{NaNH}_2}]{\text{Na or}} \text{RO}^{\ominus}\,\text{Na}^{\oplus} \quad (+\ H_2 \text{ or } NH_3)$$

$$\text{R}'\text{-OH} \xrightarrow[\text{or HX}]{\text{PBr}_3 \text{ or SOCl}_2} \text{R}'\text{-X}$$

$$\text{RO}^{\ominus} + \text{R}'\text{-X} \xrightarrow{\text{S}_N2} \text{R-O-R}' + \text{X}^{\ominus}$$

Epoxides are usually synthesized by oxidation of an alkene with a peroxy acid such as *m*-chloroperoxybenzoic acid (MCPBA). An alternative two-step procedure for conversion of an alkene to an epoxide is via a halohydrin and involves an intramolecular displacement resembling a Williamson synthesis.

Sulfides may be prepared via a Williamson-like process. Because thiols are more acidic than alcohols, the RS⁻ nucleophiles are easily prepared. The nucleophilicity of sulfur leads to some multiple substitution such that trialkylsulfonium salts can form if excess alkyl halide is present.

$$\text{RS}^- \xrightarrow[\text{R}'\text{X}]{} \underset{\text{Sulfide}}{\text{RSR}'} \xrightarrow[\text{R}'\text{X}]{} \underset{\text{Trialkylsulfonium halide}}{[\text{RSR}'_2{}^+]\text{X}^-}$$

Topic Test 2: Preparation of Ethers, Epoxides, and Sulfides

True/False

1. The Williamson ether synthesis involves a carbocation and is therefore best for tertiary alkyl halides.

2. An S_N2 displacement on an alkyl halide by an alkylthio anion will yield a sulfide.

Multiple Choice

3. Which ether below could be prepared cleanly by alcohol dehydration?
   a. 1-Methoxybutane
   b. 2-Methoxybutane
   c. Dibutyl ether
   d. All of the above
   e. None of the above

4. Which reaction below is likely to produce an epoxide?
   a. 1-Bromo-2-propanol treated with dilute aqueous hydroxide
   b. *trans*-3-Hexene treated with MCPBA
   c. Cyclopentene treated with Br_2, H_2O followed by dilute aqueous hydroxide
   d. All of the above
   e. None of the above

Short Answer

5. Show how one could prepare 1-ethoxy-1-methylcyclopentane by alkoxymercuration with ethanol and any other needed reagents.

6. Show how one could prepare 1-ethoxy-1-methylcyclopentane using the Williamson ether synthesis.

Topic Test 2: Answers

1. **False.** The S_N2 mechanism operative in the Williamson synthesis does not involve a carbocation and works best for less substituted alkyl halides (methyl and primary).

2. **True.** This is a good general method for RSR′ preparation.

3. **c.** Dibutyl ether. This is the only symmetrical ether listed. It can be prepared by heating 1-butanol with sulfuric acid.

4. **d.** All of the above. Reactions in responses a and c are via a halohydrin.

5. As usual, working backward is the easiest way to reason the problem out. The alkoxymercuration that formed this ether involved ethanol addition (Markovnikov) across an alkene. Either of two possible alkenes would lead to the desired product

6. Although on paper it appears that there are two possible alkoxide + alkyl halide combinations that would yield the desired ether, the mechanistic requirements of the S_N2 reaction limit our choice to the more substituted alkoxide and the less substituted alkyl

halide. If one attempted to use the alternative combination an elimination reaction would predominate.

TOPIC 3: REACTIONS OF ETHERS, EPOXIDES, AND SULFIDES

KEY POINTS

✓ *What is the mechanism and regiochemistry of HX ether cleavage?*

✓ *What is the mechanism and regiochemistry of epoxide ring opening in acid?*

✓ *What is the mechanism and regiochemistry of epoxide ring opening in base?*

✓ *With what reagents and to what products are sulfides oxidized?*

There are relatively few reactions of ethers. An ether can be cleaved with HI or HBr to yield an alcohol and an alkyl halide. The mechanism is normally protonation of the oxygen followed by S_N2 attack on the least hindered alkyl group. This mechanism explains the observed regiochemistry, that is, the halogen nucleophile ends up on the least substituted of the ether carbons and the most substituted ether carbon retains the oxygen to become an alcohol.

The situation is somewhat more complex for asymmetrically substituted epoxides. The regiochemistry changes depending on whether the reaction medium is acidic or basic. Under alkaline conditions, the nucleophile attacks the least hindered epoxide carbon and the oxygen becomes an alcohol OH on the more substituted carbon. In acid, the opposite regiochemistry is observed. In both cases the nucleophile attacks an epoxide carbon from the "back side," which means the nucleophile and the hydroxyl are **anti** as shown on the next page.

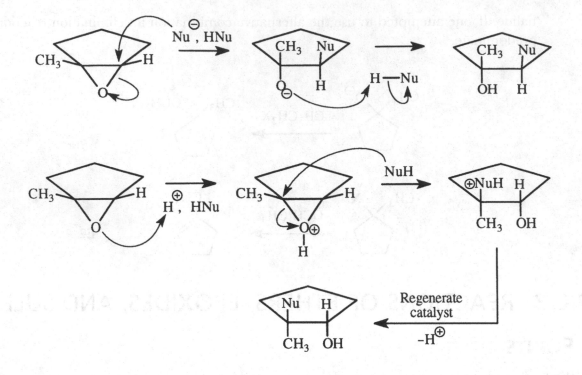

Besides acting as a nucleophile to make a sulfonium salt as shown in Topic 2, a sulfide will also undergo gentle oxidation with one equivalent of hydrogen peroxide to yield a sulfoxide or more vigorous oxidation with a second equivalent of peroxide to yield a sulfone.

$$R-S-R' \xrightarrow{H_2O_2} R-\overset{\displaystyle O}{\underset{\displaystyle }{S}}-R' \xrightarrow{H_2O_2} R-\overset{\displaystyle O}{\underset{\displaystyle O}{S}}-R'$$

Sulfide   Sulfoxide   Sulfone

Topic Test 3: Reactions of Ethers, Epoxides, and Sulfides

True/False

1. If an unsymmetrical ether is cleaved with HX, the halide will normally be attached to the most substituted carbon according to Markovnikov's rule.

2. In an acidic medium, an epoxide ring opening will generally yield a product in which the nucleophile is on the least substituted epoxide carbon.

Multiple Choice

3. Cleavage of the cyclic ether tetrahydrofuran with HI will likely produce
   a. ethyl iodide and ethanol.
   b. 4-iodo-1-butanol.
   c. 1,4-diiodobutane.
   d. 1,4-butanediol.
   e. None of the above

4. Treating propene with MCPBA followed by methanol with an acid catalyst will yield
   a. 1,2-propanediol.
   b. 1,2-dimethoxypropane.
   c. 1-methoxy-2-propanol.
   d. 2-methoxy-1-propanol.
   e. None of the above

Short Answer

5. Provide a structural formula for the missing organic compound.

6. What products would you expect from the gentle oxidation of dibutyl sulfide with one equivalent of hydrogen peroxide or more vigorous oxidation with a second equivalent of hydrogen peroxide.

Topic Test 3: Answers

1. **False.** The mechanism favors the halide to attach onto the least substituted alkyl group because the approach is least hindered. This process is unrelated to Markovnikov's rule for predicting regiochemistry of alkene addition reactions.

2. **False.** In acidic conditions the nucleophile normally ends up on the most substituted carbon.

3. **b.** 4-iodo-1-butanol ($HOCH_2CH_2CH_2CH_2I$). Except that the resulting alkyl halide and alcohol are tethered, this product is not unusual and is what one would expect from an ether cleavage by HI. Because the starting ether is symmetrical, there is no regiochemistry to consider.

4. **d.** 2-methoxy-1-propanol. The first step is epoxidation of the alkene and the second step opens the epoxide under acidic conditions to put the nucleophile on the more substituted carbon.

5.

   The acetylide anion is the nucleophile and it is alkaline. We expect the nucleophile on the least hindered side in the product.

6. The first equivalent of H_2O_2 would produce the sulfoxide, $(CH_3CH_2CH_2CH_2)_2SO$. The second equivalent would convert that to a sulfone, $(CH_3CH_2CH_2CH_2)_2SO_2$.

The strength and durability of **epoxy resins** lead to numerous commercial applications. Epoxide ring opening reactions are essential in the preparation of these adhesives. For the most common of these, a polymer of relatively low molecular weight, called a **prepolymer**, is prepared from **epichlorohydrin** and **bisphenol A**. Both these starting materials are bifunctional (i.e., they have two sites of chemical reactivity). The two nucleophilic OH groups of bisphenol A can attack either the primary alkyl chloride or the epoxide ring of epichlorohydrin. In a subsequent transformation, this prepolymer is cross-linked using a **curing agent** or **hardener** such as the amine shown.

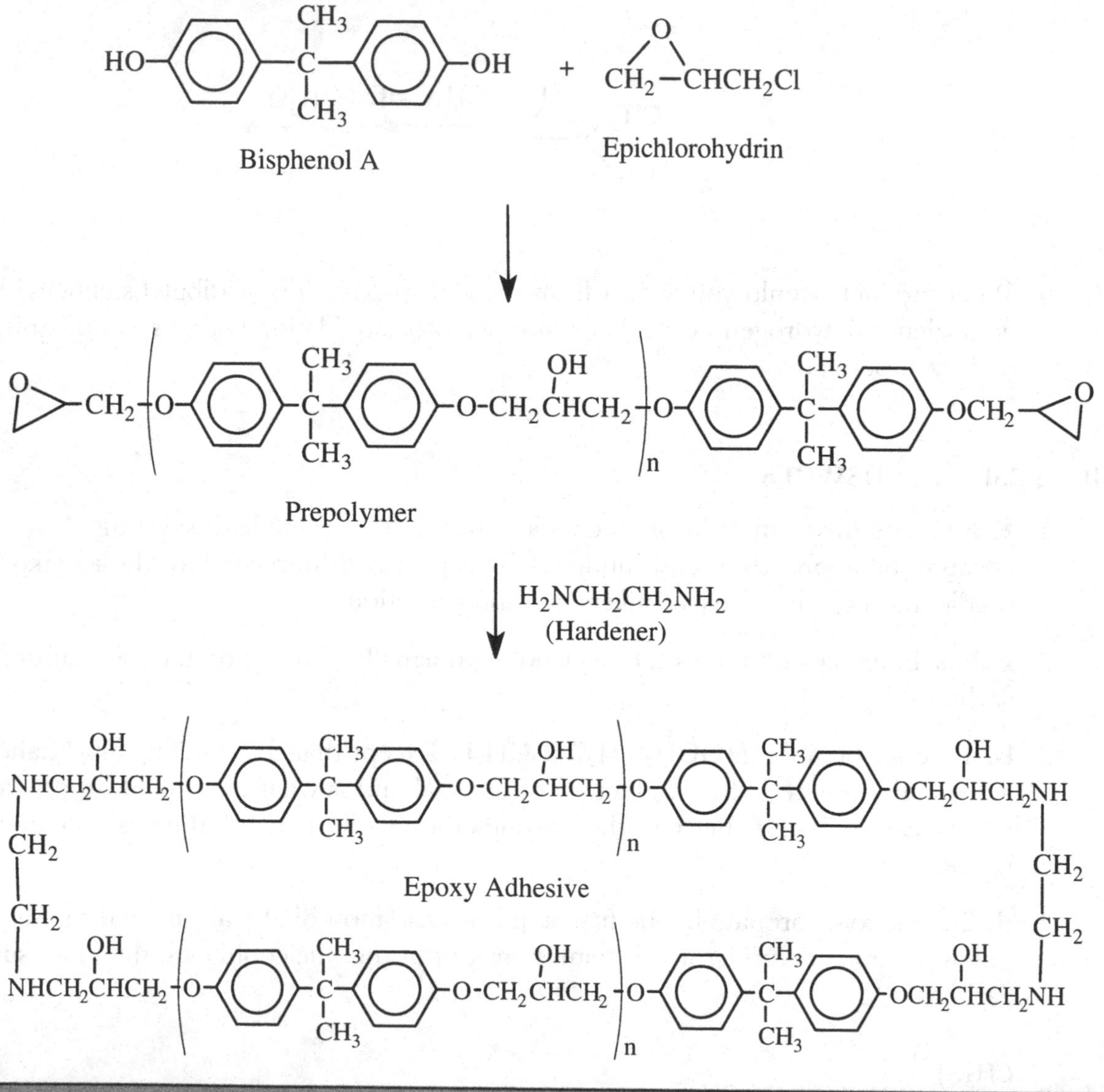

Bisphenol A   Epichlorohydrin

Prepolymer

$H_2NCH_2CH_2NH_2$
(Hardener)

Epoxy Adhesive

DEMONSTRATION PROBLEM

Show the reagents and conditions one could use to prepare 1,2-diethyoxybutane from 1-butene.

Solution

There is more than one possible correct combination of steps, but all obvious and reasonable strategies involve epoxidation of the alkene, ring opening with an oxygen nucleophile, and a Williamson ether synthesis. Some possible reagent combinations are shown below.

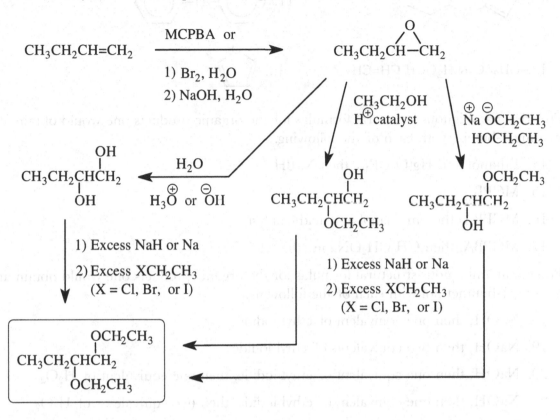

Chapter Test
True/False

1. Ethers are generally more water soluble than alcohols of similar size.

2. Dibutyl ether can be made by alkoxymercuration starting with 1-butene and 1-butanol.

3. The Williamson ether synthesis is good strategy for preparation of $(CH_3)_3COC(CH_3)_3$.

4. Ether cleavage with HI produces an alcohol in which the ether oxygen becomes an alcohol on the more substituted carbon of the original ether.

5. The key step in the Williamson ether synthesis is an S_N2 reaction.

6. Epoxide ring opening under alkaline conditions leads to a product in which the nucleophile is attached to the least substituted of the epoxide carbons.

Short Answer

Provide structural formulas for the following.

7. *trans*-1,4-Diethoxycyclohexane

8. Dibenzyl ether

9. Tetrahydrofuran

Name the following.

10.

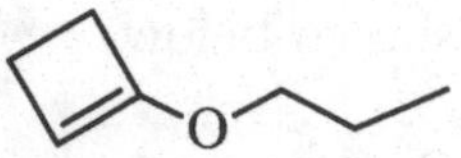

11.

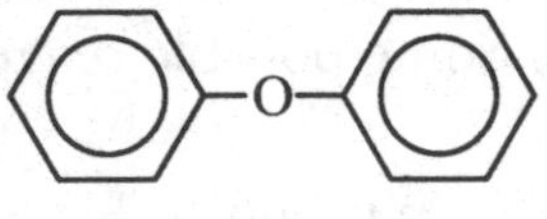

12. $CH_2=CHCH_2OCH_2CH=CH_2$

13.

Provide unambiguous structural formulas for the organic products one would obtain after treating 1-butene with each of the following.

14. Ethanol and $Hg(O_2CF_3)_2$, then $NaBH_4$

15. MCPBA

16. MCPBA, then methanol and acid catalyst

17. MCPBA, then CH_3CH_2ONa in ethanol

Provide unambiguous structural formulas for the organic products one would obtain after treating 1-butanethiol with each of the following.

18. NaOH, then one equivalent of ethyl iodide

19. NaOH, then two equivalents of ethyl iodide

20. NaOH, then one equivalent of ethyl iodide, then one equivalent of H_2O_2

21. NaOH, then one equivalent of ethyl iodide, then two equivalents of H_2O_2

22. Show how one could convert 1-pentene into an epoxide via a halohydrin. More than one step will be required.

23. Show how one could prepare t-butyl propyl ether via a Williamson ether synthesis and using alcohols as the only carbon source. More than one step will be required.

24. What product would be formed in the following reaction?

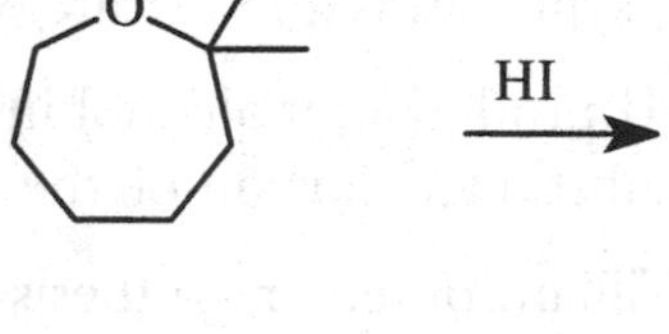

Chapter Test: Answers

1. **False**

2. **False**

3. **False**

4. **True**

5. **True**

6. **True**

7. 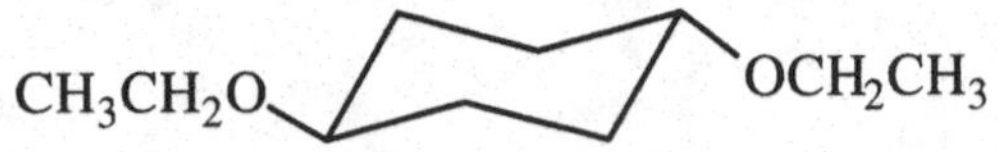 CH_3CH_2O—[cyclohexane]—OCH_2CH_3

8. 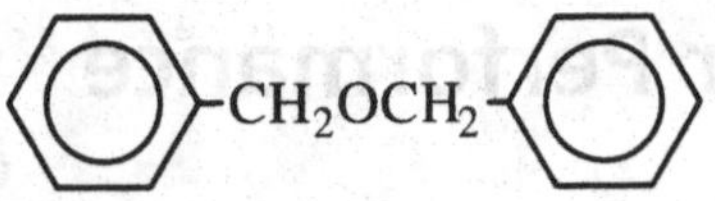 [benzene]—CH_2OCH_2—[benzene]

9.  [tetrahydrofuran structure with O]

10. 1-Propoxycyclobutene

11. Diphenyl ether

12. Diallyl ether

13. Furan

14. 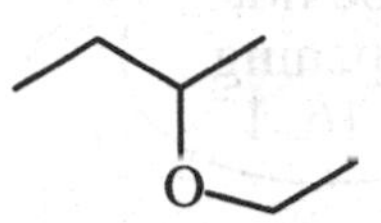 [structure with O—ethyl ether group]

15. 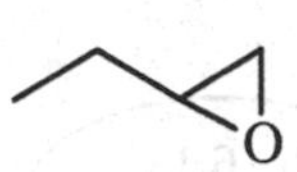 [propylene oxide epoxide structure with O]

16. 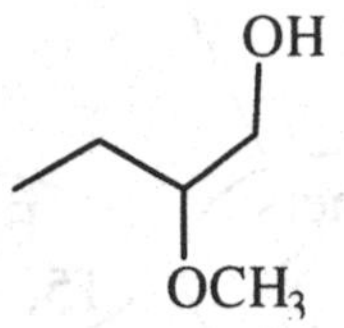 [structure with OH and OCH_3]

17. 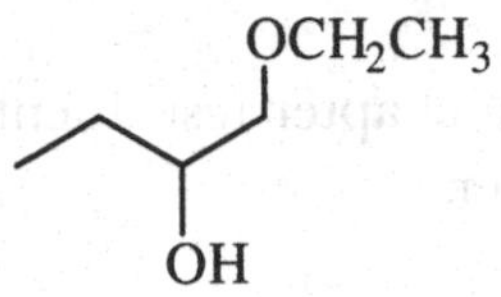 [structure with OCH_2CH_3 and OH]

18.  [structure with S]

19. $[CH_3CH_2CH_2\overset{\oplus}{S}(CH_2CH_3)_2]\,\overset{\ominus}{I}$

20. 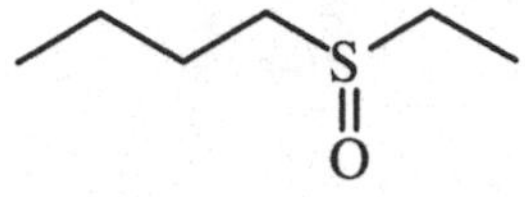 [structure with S=O, sulfoxide]

21. 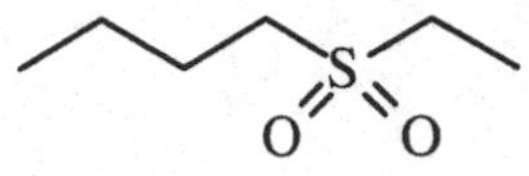 [structure with SO_2 sulfone, O=S=O]

22. 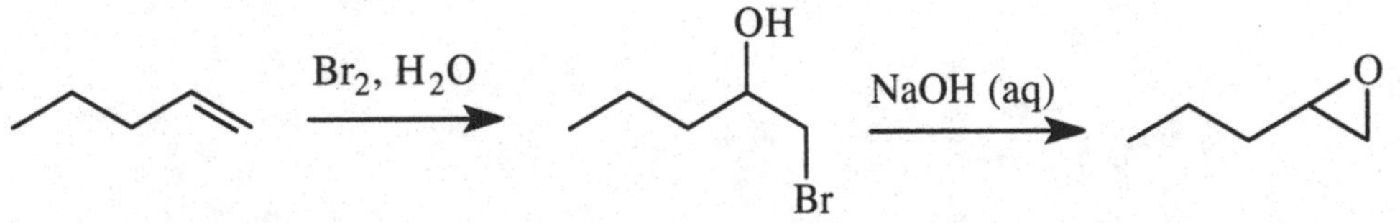 [pent-4-ene structure] $\xrightarrow{\text{Br}_2,\ \text{H}_2\text{O}}$ [2-pentanol with Br, OH] $\xrightarrow{\text{NaOH (aq)}}$ [epoxide structure with O]

23. 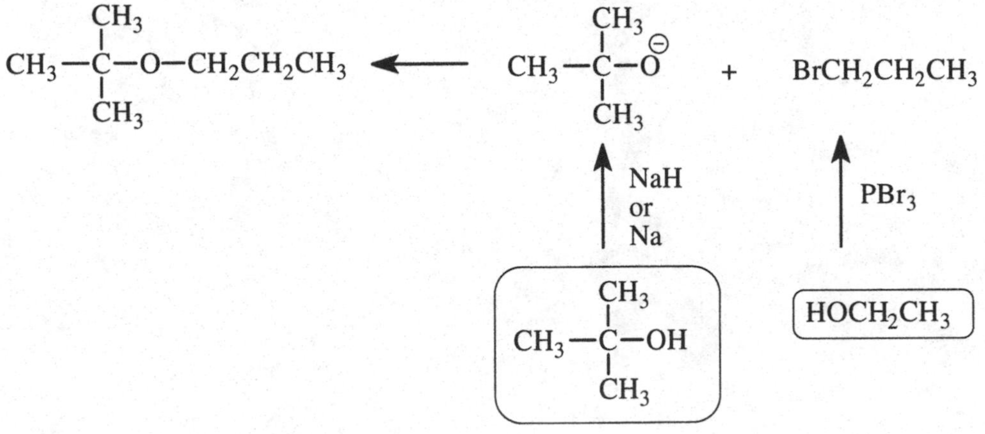
$$CH_3\text{—}\underset{\underset{CH_3}{|}}{\overset{\overset{CH_3}{|}}{C}}\text{—}O\text{—}CH_2CH_2CH_3 \longleftarrow CH_3\text{—}\underset{\underset{CH_3}{|}}{\overset{\overset{CH_3}{|}}{C}}\text{—}\overset{\ominus}{O} \quad + \quad BrCH_2CH_2CH_3$$

$$\uparrow \begin{array}{c}\text{NaH}\\\text{or}\\\text{Na}\end{array} \qquad \uparrow \text{PBr}_3$$

$$\boxed{CH_3\text{—}\underset{\underset{CH_3}{|}}{\overset{\overset{CH_3}{|}}{C}}\text{—}OH} \qquad \boxed{HOCH_2CH_3}$$

24. 7-Iodo-2-methyl-2-heptanol

Check Your Performance

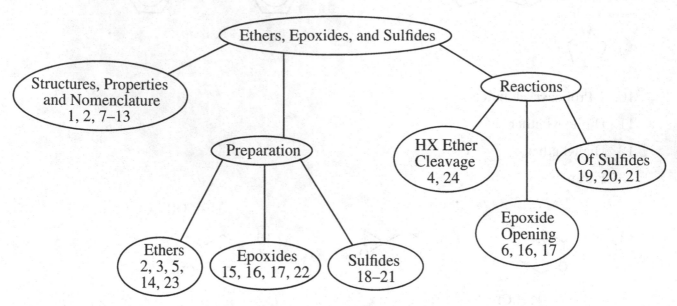

Note the number of questions in each grouping that you got wrong on the chapter test. Identify areas where you need further review and go back to relevant parts of this chapter.

CHAPTER 5

Aldehydes and Ketones

Carbonyl compounds are abundant in nature and important in the study of organic chemistry. This is the first chapter dedicated to compounds containing the C=O linkage and where we survey some of the chemistry of the simplest carbonyl compounds. Aldehydes and ketones are so similar in their patterns of reactivity that they can conveniently be studied together.

ESSENTIAL BACKGROUND

- **Equilibrium and Le Châtelier's principle (general chemistry)**
- **Ozonolysis of alkenes**
- **Hydration of alkynes and tautomerism of enols to carbonyl compounds**
- **Conjugation**
- **Friedel-Crafts acylation (Chapter 1)**
- **Hydride reductions aldehydes and ketones (Chapter 3)**
- **Alcohols from Grignard reagents and aldehydes or ketones (Chapter 3)**
- **Oxidation of alcohols (Chapter 3)**

TOPIC 1: STRUCTURE AND NOMENCLATURE

KEY POINTS

✓ *What are the structures of aldehydes and ketones?*

✓ *How are aldehydes and ketones named?*

Aldehydes have the general structure RCH=O where at least one hydrogen is bound directly to the carbonyl carbon and the other side of the carbonyl is attached to hydrogen or carbon (R=H, alkyl, aryl, etc.). Simple aldehydes are named by finding the longest chain containing the carbonyl and replacing the "ane" suffix of the corresponding alkane parent with the suffix "al." The carbonyl carbon is defined as C1, and substituents are included as prefixes with numbers as locators. If R is cyclic, the name is derived from the corresponding parent molecule with the suffix "carbaldehyde." Some widely used common names for aldehydes should be known and are illustrated in **Table 5.1**.

Table 5.1 Names and Structures of Representative Aldehydes

STRUCTURE	SYSTEMATIC IUPAC NAME	COMMON NAME
H–C–H (with O double bond)	Methanal	Formaldehyde
CH₃CH (with O double bond)	Ethanal	Acetaldehyde
CH₃CH₂CH (with O double bond)	Propanal	Propionaldehyde
CH₃CH₂CH₂CH (with O double bond)	Butanal	Butyraldehyde
CH₃CH₂CH₂CH₂CH (with O double bond)	Pentanal	Valeraldehyde
benzene ring–C–H (with O double bond)	Benzenecarbaldehyde	Benzaldehyde
branched chain–C–H (with O double bond)	2-Ethylbutanal	
CH₃CH₂CHCHCH with OH and O double bond, CH₃	3-Hydroxy-2-methylpentanal	
cyclohexane ring with C=O–H	Cyclohexanecarbaldehyde	
cyclopentene ring with C–H (O double bond)	1-Cyclopentenecarbaldehyde	

Ketones have a carbonyl bound directly to two carbons. Systematic names of ketones are determined by finding the longest chain or ring that contains the carbonyl and replacing the "e" at the end of the corresponding alkane name with the suffix "one." Numbers are assigned such that the carbonyl gets the lowest number possible. Some widely used common names for ketones should be known and are illustrated in **Table 5.2**.

Topic Test 1: Structure and Nomenclature

True/False

1. Systematic IUPAC aldehyde names generally end in "ol."

2. Propanone and acetone are two names for the same thing.

Table 5.2 Names and Structures of Representative Ketones

STRUCTURE	NAME(S)
CH_3CCH_3 (O double bond)	Propanone (acetone)
$CH_3CH_2CCH_3$ (O double bond)	Butanone (methyl ethyl ketone)
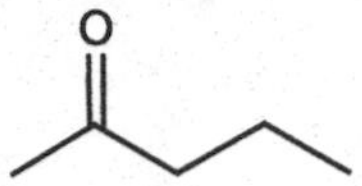 (structure)	2-Pentanone
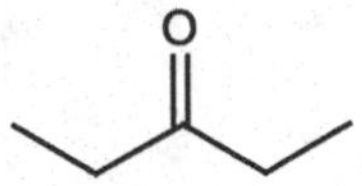 (structure)	3-Pentanone
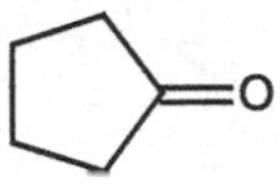 (structure)	Cyclopentanone
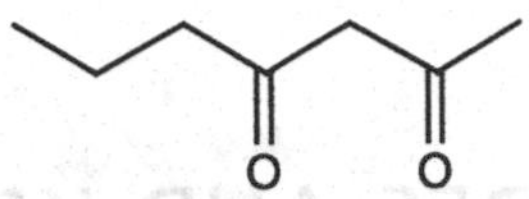 (structure)	2,4-heptanedione
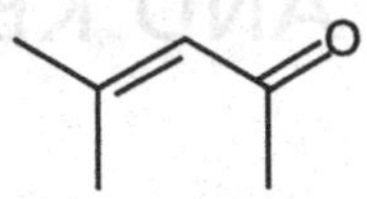 (structure)	4-Methyl-3-penten-2-one
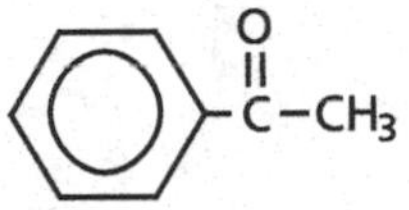 (structure) $C-CH_3$	Acetophenone
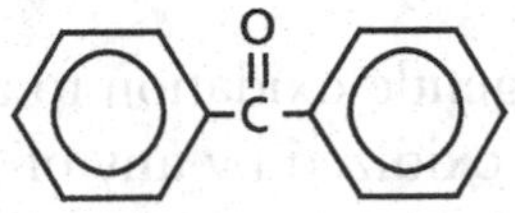 (structure)	Benzophenone

Short Answer

Provide unambiguous structural formulas for the compounds named.

3. Phenylacetaldehyde

4. (*E*)-4-decene-3-one

Name the following compounds

5. CH_3CHCH_2CH (O double bond)
   $\quad\ \ |$
   $\quad\ \ CH_3$

6. 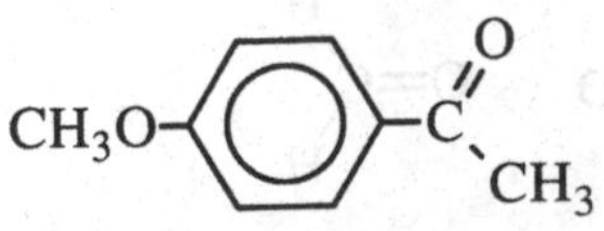 CH_3O—(benzene ring)—C (O double bond)
   $\quad\quad\quad\quad\quad\quad\quad\quad CH_3$

Topic Test 1: Answers

1. **False.** They usually end in "al." The "ol" suffix is used for alcohols. Be cautious here. Some people pronounce "ethanal" and "ethanol" in a way that makes them difficult to distinguish.

2. **True.** The systematic IUPAC name is propanone; however, acetone is the more widely used name.

3.

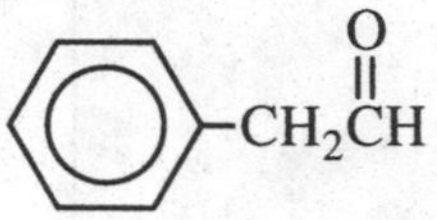

4. 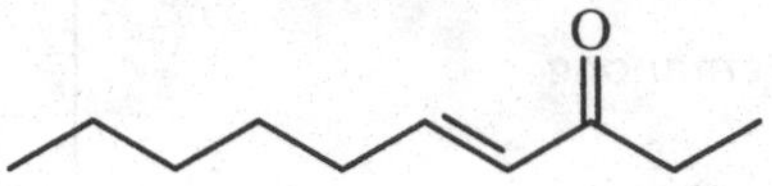   This could also be called *trans*-4-decen-3-one

5. 3-Methylbutanal

6. *p*-Methoxyacetophenone

TOPIC 2: PREPARATIONS OF ALDEHYDES AND KETONES

KEY POINTS

✓ *What reactions from previous chapters yield aldehydes and ketones?*

✓ *How are ketones made from Gilman reagents?*

Recall from Chapter 3 that primary alcohols can undergo gentle oxidation to aldehydes with pyridinium chlorochromate (PCC). Secondary alcohols are oxidized by any of several reagents (Jones', $KMnO_4$, PCC, etc.) to yield ketones.

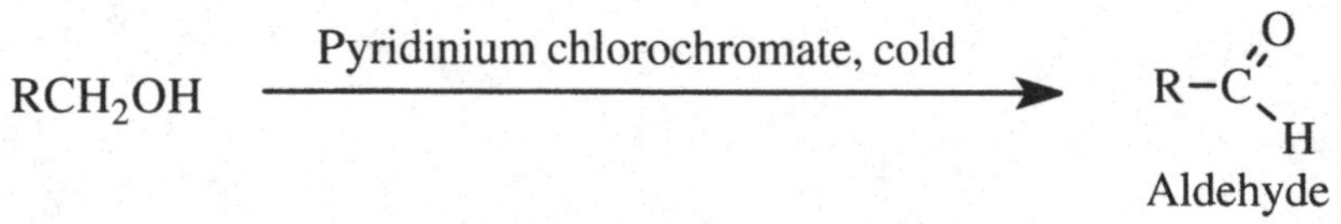

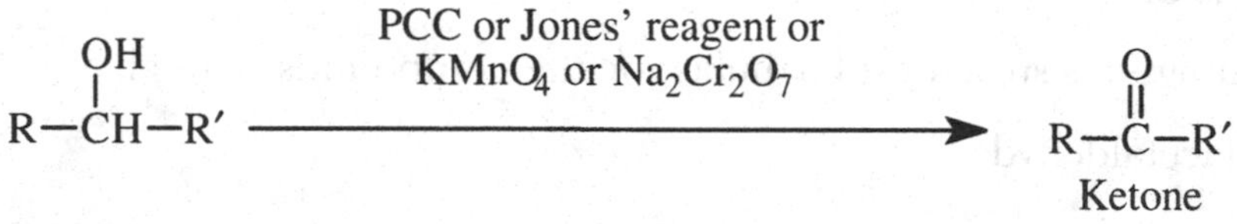

Ozonolysis of alkenes leads to aldehydes and ketones.

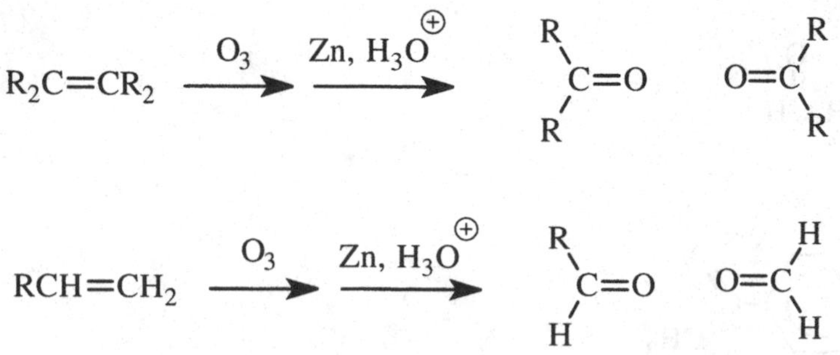

Hydration of alkynes (**oxymercuration** or **hydroboration**) yields aldehydes and ketones via the enol tautomers.

$$R-C\equiv C-H \xrightarrow[\substack{HgSO_4 \\ H_2SO_4}]{H_2O} \left[R-\underset{\underset{\text{Markovnikov enol}}{}}{\overset{\overset{OH}{|}}{C}}=CH_2 \right] \rightleftharpoons R-\underset{\underset{\text{A methyl ketone}}{}}{\overset{\overset{O}{\|}}{C}}-CH_3$$

Terminal alkyne

$$R-C\equiv C-R \xrightarrow[\substack{HgSO_4 \\ H_2SO_4}]{H_2O} \left[R-\overset{\overset{OH}{|}}{C}=CHR \right] \rightleftharpoons R-\underset{\underset{\text{A ketone}}{}}{\overset{\overset{O}{\|}}{C}}-CH_2R$$

Symmetrical alkyne

$$R-C\equiv C-R' \xrightarrow[\substack{HgSO_4 \\ H_2SO_4}]{H_2O} \left[\begin{array}{c} R-\overset{\overset{OH}{|}}{C}=CHR' \\ + \\ R-CH=\overset{\overset{OH}{|}}{C}R' \end{array} \right] \rightleftharpoons \begin{array}{c} R-\overset{\overset{O}{\|}}{C}-CH_2R' \\ + \\ R-CH_2-\overset{\overset{O}{\|}}{C}R' \end{array}$$

Unsymmetrical alkyne

Two different ketones

$$R-C\equiv C-H \xrightarrow{BH_3} \left[\begin{array}{c} H \quad B\diagdown \\ | \quad | \\ R-\overset{}{C}-\overset{}{CH} \\ | \quad | \\ H \quad B\diagup \end{array} \right] \xrightarrow[H_2O_2]{OH^{\ominus}} \left[R-CH_2-\overset{\overset{}{CH}}{\underset{\underset{OH}{|}}{\overset{\overset{OH}{|}}{}}} \right] \rightleftharpoons \begin{array}{c} \overset{O}{\|} \\ RCH_2CH \\ + \ H_2O \end{array}$$

Finally, recall from Chapter 1 that aromatic ketones (phenones) can be prepared by the **Friedel-Crafts acylation** provided the aromatic ring is not strongly deactivated.

Acyl chlorides such as those used in the Friedel-Crafts reaction can also be coupled to Gilman reagents to yield ketones.

Topic Test 2: Preparations of Aldehydes and Ketones

True/False

1. Hydroboration of 1-pentyne with BH_3 followed by H_2O_2, OH^- yields 2-pentanone.

2. Nitrobenzene will undergo a Friedel-Crafts acylation to give an aromatic ketone.

Multiple Choice

3. Which reaction below will yield a ketone?
   a. Treating a secondary alcohol with PCC
   b. Friedel-Crafts acylation of benzene
   c. Treating an alkyne with H_2O, $HgSO_4$, H_2SO_4
   d. All of the above
   e. None of the above

4. Which reaction below will yield an aldehyde?
   a. Ozonolysis of 2,3-dimethyl-2-butene
   b. Treating 1-propanol with PCC
   c. Treating 1-propanol with Jones' reagent
   d. All of the above
   e. None of the above

Short Answer

Provide unambiguous structural formulas for the missing organic compounds.

5.

6.

Topic Test 2: Answers

1. **False.** The two-step hydroboration strategy leads to a non-Markovnikov enol that tautomerizes to pentanal. To make 2-pentanone from 1-pentyne, use H_2O, $HgSO_4$, H_2SO_4.

2. **False.** The strong electron-withdrawing nitro group deactivates the ring so much that it cannot participate in a Friedel-Crafts reaction.

3. **d.** All of the above

4. **b.** The other choices all lead to nonaldehyde products. Treating 1-propanol with PCC leads to propanol, CH_3CH_2CHO. Ozonolysis of 2,3-dimethyl-2-butene yields two equivalents of acetone. 1-Propanol would oxidize all the way to a carboxylic acid ($CH_3CH_2CO_2H$) if treated with Jones' reagent.

5. $\overset{O}{\overset{\|}{H C}}CH_2CH_2CH_2CH_2\overset{O}{\overset{\|}{C}}CH_3$

6.

TOPIC 3: NUCLEOPHILIC ADDITION

KEY POINTS

✓ *What is the general form of a nucleophilic addition to a carbonyl?*

✓ *What is the relative reactivity of aldehydes and ketones toward addition?*

✓ *What addition reactions from previous chapters convert carbonyls to alcohols?*

✓ *What is a hydrate and how does it form?*

✓ *What are cyanohydrins and how are they prepared?*

The carbonyl groups of aldehydes and ketones frequently undergo addition reactions. The electronegative oxygen of a carbonyl polarizes the double bond toward oxygen such that the more negative end of the adding reagent attaches on carbon. The process can also be acid catalyzed in which case initial protonation of the carbonyl oxygen makes the carbonyl carbon more susceptible to nucleophilic attack. Notice that the acid is not consumed in the overall process.

Aldehydes are generally a little more reactive than ketones for both steric and electronic reasons. We have seen examples of additions to carbonyls in previous chapters. Recall from Chapter 3 that aldehydes and ketones will react with hydride (from $LiAlH_4$ or $NaBH_4$) or with Grignard reagents (essentially R^- anions) followed by acid to yield alcohols.

$$\text{R}-\overset{\overset{\textstyle O}{\|}}{\text{C}}-\text{R}' \quad \xrightarrow[\substack{\text{NaBH}_4 \text{ or} \\ \text{LiAlH}_4}]{"\text{H}^{\ominus}"} \quad \text{R}-\overset{\overset{\textstyle O^{\ominus}}{|}}{\underset{\underset{\textstyle H}{|}}{\text{C}}}-\text{R}' \quad \xrightarrow{\text{H}_3\text{O}^{\oplus}} \quad \text{R}-\overset{\overset{\textstyle OH}{|}}{\underset{\underset{\textstyle H}{|}}{\text{C}}}-\text{R}'$$

$$\text{R}-\overset{\overset{\textstyle O}{\|}}{\text{C}}-\text{R}' \quad (\text{R, R}' = \text{H, alkyl, aryl})$$

$$\xrightarrow[\text{R}''\text{MgX}]{"\text{R}^{\ominus}"} \quad \text{R}-\overset{\overset{\textstyle O^{\ominus}}{|}}{\underset{\underset{\textstyle R''}{|}}{\text{C}}}-\text{R}' \quad \xrightarrow{\text{H}_3\text{O}^{\oplus}} \quad \text{R}-\overset{\overset{\textstyle OH}{|}}{\underset{\underset{\textstyle R''}{|}}{\text{C}}}-\text{R}'$$

Two other simple additions to aldehydes and ketones are the formation of **hydrates** and **cyanohydrins**, which formally amount to addition of water or HCN across the carbonyl, respectively. Hydrate formation is a reversible equilibrium that, in most cases, favors the carbonyl compound.

$$\underset{\substack{\text{Aldehyde} \\ \text{or Ketone}}}{\text{R}-\overset{\overset{\textstyle O}{\|}}{\text{C}}-\text{R}'} \;+\; \text{H}_2\text{O} \;\rightleftharpoons\; \underset{\text{Hydrate}}{\text{R}-\overset{\overset{\textstyle OH}{|}}{\underset{\underset{\textstyle OH}{|}}{\text{C}}}-\text{R}'}$$

$$\text{R}-\overset{\overset{\textstyle O}{\|}}{\text{C}}-\text{R}' \quad \xrightarrow{\text{HCN}} \quad \underset{\text{Cyanohydrin}}{\text{R}-\overset{\overset{\textstyle OH}{|}}{\underset{\underset{\textstyle \overset{\textstyle |||}{N}}{\overset{\textstyle C}{|}}}{\text{C}}}-\text{R}'}$$

Topic Test 3: Nucleophilic Addition

True/False

1. The product of a nucleophilic addition to a carbonyl usually has the nucleophile attached to the carbonyl oxygen.

2. Acetone will react with CH_3MgBr and then H_3O^+ to yield *t*-butyl alcohol.

Multiple Choice

3. Which of the following would react fastest with a given nucleophile?
   a. Methanal (formaldehyde)
   b. Ethanal (acetaldehyde)
   c. Propanone (acetone)
   d. 2-Butanone
   e. These all react at the same rate.

4. Which reaction below is an example of nucleophilic addition to a carbonyl?
   a. Ozonolysis of alkenes

b. Hydration of alkynes

c. Reduction of ketones with NaBH$_4$ and then H$_3$O$^+$

d. Oxidation of a primary alcohol with PCC

e. None of the above

Short Answer

5. Write a balanced chemical equation showing the equilibrium between cyclopentanone and its hydrate. Indicate which organic compound you expect to be present in the larger amount.

6. Write a chemical reaction scheme showing how one could convert benzaldehyde into a cyanohydrin.

Topic Test 3: Answers

1. **False.** The polarity of the C=O is such that the nucleophile attaches to the carbon.

2. **True.** Reactions of this type were discussed in detail in Chapter 3 and then again in this topic. This is an example of nucleophilic addition to a carbonyl.

3. **a.** Methanal (formaldehyde) reacts fastest. Steric and electronic effects make a carbonyl more reactive when it is attached to a hydrogen(s) rather than an alkyl group(s). Methanal is the most reactive of all the aldehydes, and aldehydes are generally more reactive than ketones.

4. **c.** Reduction of ketones with NaBH$_4$ and then H$_3$O$^+$. In the first step, NaBH$_4$ delivers a hydride to the carbonyl carbon, generating a secondary alkoxide that is protonated at oxygen when the acid is subsequently added.

5.

favored

6.

TOPIC 4: MORE ADDITIONS TO CARBONYLS

KEY POINTS

✓ *What are hemiacetals and acetals and how are they prepared?*

✓ *What are imines (Schiff bases) and how are they prepared?*

✓ *What are enamines and how are they prepared?*

✓ *What are Wittig reagents and how are they prepared?*

✓ *How do Wittig reagents react with aldehydes and ketones?*

Alcohols can add to the carbonyl groups of aldehydes or ketones to yield **hemiacetals**. The reaction is often acid catalyzed. If a second equivalent of alcohol is available, the hemiacetal reacts further to yield an **acetal**. Note that in an acetal or hemiacetal, two oxygens flank the original carbonyl carbon. Notice also that inspection of that central carbon in the product (to see if it has H attached to it) can reveal whether the original carbonyl was that of an aldehyde or a ketone. Acetal and hemiacetal formation is reversible, and the equilibrium can be manipulated according to Le Châtelier's principle. If excess water and acid are added to an acetal, the resulting hydrolysis yields an aldehyde or ketone.

When a monosubstituted ammonia, $Y—NH_2$, reacts with an aldehyde or ketone the resulting product is an **imine** (sometimes called a **Schiff base**). The transformation involves nitrogen attack at the carbonyl to yield an unstable intermediate addition product that spontaneously dehydrates to the imine. The substituent Y can be many different things, some of which are shown in **Table 5.3** along with the names and structures of the products.

If a secondary amine adds to an aldehyde or ketone, the resulting intermediate addition product cannot dehydrate to an imine because there is no H on the nitrogen. Provided there is at least

Table 5.3 Reagents, Structures, and Names of Schiff Bases

REAGENT, $Y—NH_2$	PRODUCT TYPE
Primary amine, RNH_2	Imine
Hydrazine, NH_2NH_2	Hydrazone
2,4-Dinitrophenylhydrazine, O_2N—⬡(NO_2)—$NHNH_2$	2,4-Dinitrophenylhydrazone
Hydroxylamine, $HONH_2$	Oxime
Semicarbazide, $H_2NCNHNH_2$	Semicarbazone

one H on the carbon adjacent to the original carbonyl, the subsequent dehydration involves proton loss from a carbon to yield an **enamine** (ene + amine).

Aldehydes and ketones can be converted to alkenes through the use of **phosphorus ylides** commonly called **Wittig reagents**. Wittig reagents themselves are made in two steps from methyl, primary, or secondary alkyl halides (S_N2) that are reacted with triphenylphosphine (PPh_3, Ph = phenyl) and then irreversibly deprotonated by a strong base such as butyl lithium. A limitation of this method of alkene synthesis is that when *cis/trans* isomers can be formed, a mixture of products results. The net transformation is carbonyl to alkene; thus, the reaction resembles the reverse of alkene ozonolysis.

Topic Test 4: More Additions to Carbonyls

True/False

1. Acetal formation is reversible.

2. Primary amines, RNH_2, generally add to aldehydes to give imines.

Multiple Choice

3. Which compound below would react with 3-pentanone to form an enamine?
   a. NH_3

b. CH_3NH_2

c. $(CH_3)_2NH$

d. All of the above

e. None of the above

4. Which of the following is a Wittig reagent?

a. Ph_3P

b. $[Ph_3PCH_3]^+ I^-$

c. $Ph_3P{=}CH_2$

d. All of the above

e. None of the above

Short Answer

5. Draw the hemiacetal and acetal that form when cyclohexanone undergoes an acid-catalyzed reaction with one equivalent or two equivalents of ethanol, respectively.

6. Show how one could make the appropriate Wittig reagent and then react it with 3-pentanone to yield 3-ethyl-2-pentene.

Topic Test 4: Answers

1. **True.** The alcohol and carbonyl compound can easily be regenerated by adding water (hydrolysis) and/or removing excess alcohol (Le Châtelier's principle).

2. **True.**

3. **c.** $(CH_3)_2NH$. This is a secondary amine (i.e., two groups on nitrogen) and will yield an enamine when reacted with 3-pentanone. The other two reagents would yield imines if reacted with 3-pentanone.

4. **c.** $Ph_3P{=}CH_2$ is the only phosphorus ylide. Such ylides are often represented by a different resonance form with a negative charge on the methylene carbon, a positive charge on the phosphorus, and a single bond between them.

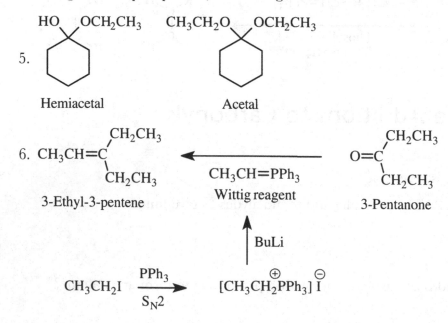

5.

Hemiacetal   Acetal

6.

3-Ethyl-3-pentene   Wittig reagent   3-Pentanone

BuLi

TOPIC 5: OTHER REACTIONS

KEY POINTS

✓ *What reagents and/or conditions will reduce a carbonyl to a methylene (CH₂)?*

✓ *What reagents and/or conditions will oxidize an aldehyde?*

✓ *How are Greek letters used to refer to positions in a carbonyl-containing compound?*

✓ *What is the Cannizzaro reaction?*

There are several ways to reduce the carbonyl of an aldehyde or ketone to a methylene group. The **Wolff-Kishner** reduction requires hydrazine and potassium hydroxide. An alternative method is called the **Clemmensen reduction**, and it uses zinc-mercury amalgam in concentrated hydrochloric acid. Aromatic ketones (phenones) undergo reduction with hydrogen and palladium catalyst. This last method is not general for all aldehydes and ketones but is limited to those in which the carbonyl is conjugated with the aromatic ring.

(Ar = Phenyl or substituted phenyl)

Aldehydes undergo oxidation to carboxylic acids under a variety of conditions, including Jones' reagent, permanganate, and dichromate. These reagents, however, will also oxidize any primary or secondary alcohols that might be present. A gentle and selective reagent for aldehyde oxidation is **Tollens' reagent** made from an aqueous ammonia solution containing silver ion. The silver ion is reduced to silver metal in the process, and if the reaction vessel is clean, the silver will deposit a mirror finish onto the inside of the flask. This is the basis for the Tollens' silver mirror test. The organic product is a carboxylate that is protonated to give the carboxylic acid upon acidification.

It is often convenient to refer to some position in a molecule as it relates to the carbonyl. One widely used system assigns Greek letters to the positions, beginning with alpha adjacent to the carbonyl, then beta, and so on. If the carbonyl is in the middle of a chain, then there can be more than one alpha position.

$$\underset{\delta\ \ \gamma\ \ \beta\ \ \alpha}{C-C-C-C}\overset{\overset{\textstyle O}{\|}}{-C-}\underset{\alpha\ \ \beta\ \ \gamma\ \ \delta}{C-C-C-C}$$

Aldehydes that have no alpha hydrogens (i.e., those with no hydrogen on the carbon next to carbonyl such as formaldehyde and benzaldehyde) can undergo an unusual reaction when subjected to vigorous conditions of potassium hydroxide and heat followed by aqueous acid. The process is called the **Cannizzaro reaction**, and it is actually a **disproportionation**, that is, an oxidation-reduction reaction in which half the starting material is oxidized and the other half is reduced. The products are a primary alcohol and a carboxylic acid.

$$\overset{\overset{\textstyle O}{\|}}{R-C-H}\ \xrightarrow[\Delta]{KOH}\ \xrightarrow{H_3O^{\oplus}}\ \overset{\overset{\textstyle O}{\|}}{R-C-OH}\ +\ R-CH_2OH$$

(R = H, aryl, or alkyl with no α hydrogen)

Topic Test 5: Other Reactions

True/False

1. Hydrogen gas with Pd catalyst in ethanol will reduce acetone to propane.

2. Benzaldehyde and formaldehyde have no alpha hydrogens.

Multiple Choice

3. What reagents and or conditions will oxidize an aldehyde to a carboxylic acid?
   a. NH_2NH_2, KOH, heat (Wolff-Kishner)
   b. Zn(Hg), H_3O^+ (Clemmensen)
   c. $Ag(NH_3)_2^+$ (Tollens'), then H_3O^+
   d. All of the above
   e. None of the above

4. Which is a characteristic of the Cannizzaro reaction?
   a. It is a disproportionation.
   b. The starting material is an aldehyde without alpha hydrogens.
   c. Half the product is reduced to a primary alcohol and half is oxidized to a carboxylate or carboxylic acid.
   d. All of the above
   e. None of the above

Short Answer

5. List three ways to convert benzaldehyde to toluene.

6. List two ways to convert benzaldehyde to benzoic acid.

Topic Test 5: Answers

1. **False.** This method of reduction works only for carbonyls of aldehydes and ketones that are conjugated with an aromatic ring.

2. **True.** Alpha hydrogens are, by definition, hydrogens on carbon adjacent to a carbonyl. Neither benzaldehyde nor formaldehyde has these.

3. **c.** $Ag(NH_3)_2^+$ (Tollens'), then H_3O^+. This is the basis for the silver mirror test. The silver is reduced and the aldehyde is oxidized. The other reagents a and b are for reducing a carbonyl of an aldehyde or ketone to a methylene.

4. **d.** All of the above

5. H_2, Pd, ethanol; Clemmensen reduction; and Wolff-Kishner reduction.

6. Tollens' silver mirror test and Jones' reagent (other correct answers are possible).

TOPIC 6: CONJUGATE ADDITION TO α,β-UNSATURATED CARBONYL COMPOUNDS

KEY POINTS

✓ *What is an α,β-unsaturated carbonyl compound?*

✓ *How are normal addition and conjugate addition defined?*

✓ *What are some normal addition and some conjugate addition nucleophiles?*

We saw in Topic 3 that addition to a carbonyl usually involves attack by a nucleophile on the carbonyl carbon. When a pi bond is in conjugation with a carbonyl, an alternative pathway of reactivity becomes available. Such structures are described as α,β-**unsaturated**, and the nucleophile can attack at the carbonyl carbon (normal addition) or at the beta position (conjugate addition) as shown.

Normal Addition

Conjugate Addition

Whether a given reaction with an α,β-unsaturated carbonyl compound will lead to a normal addition or a conjugate addition product is predictable based on the nucleophile in many cases. Grignard reagents usually lead to normal addition products. Gilman reagents (organocuprates, $LiCuR_2$) and amines usually lead to conjugate addition products.

Topic Test 6: Conjugate Addition to α,β-Unsaturated Carbonyl Compounds

True/False

1. 3-Cyclohexenone is an α,β-unsaturated carbonyl compound.

2. Conjugate addition results from attack of a nucleophile at the alpha position.

Multiple Choice

3. Which of the following reagents would you expect to yield a normal addition product in a reaction with 2-cyclohexenone?
   a. $[(CH_3CH_2)_2Cu]Li$
   b. $(CH_3CH_2)_2NH$
   c. CH_3CH_2MgBr
   d. All of the above
   e. None of the above

4. Which reagent below would you expect to give a conjugate addition product in a reaction with 2-propenal (common name = acrolein, $CH_2{=}CH{-}CH{=}O$)?
   a. Ph_2CuLi, then H_3O^+
   b. $(CH_3CH_2)_2NH$
   c. $CH_3CH_2NH_2$
   d. All of the above
   e. None of the above

Short Answer

5. Provide an unambiguous structural formula for the missing organic product.

6. Show how one could synthesize the following compound using a conjugate addition reaction.

Topic Test 6: Answers

1. **False.** There is an sp^3-hybridized carbon (carbon 2) insulating the alkene from the carbonyl. 3-Cylohexenone is not conjugated and cannot undergo conjugate addition.

2. **False.** For conjugate addition to occur the nucleophile must attack the beta position of an α,β-unsaturated carbonyl.

3. **c.** CH_3CH_2MgBr. Grignard reagents usually add at the carbonyl (normal addition), whereas Gilman reagents (a) and amines (b) usually give conjugate addition products.

4. **d.** All of the above.

5.

6.

1) $[(CH_3CH_2)_2Cu]Li$
2) $H_3O^{\oplus}$

DEMONSTRATION PROBLEM

Show the reagents and conditions one could use to carry out the transformation below. More than one step will be required.

Solution

As usual, the most reliable way to work a multistep synthesis problem like this is to analyze it backward. Notice that the starting material has 9 carbons, whereas the product contains 11 carbons. Clearly, the two additional carbons must have been introduced along the way. Because the two new carbons are not attached to one another, it seems likely that they were introduced in separate steps. The exocyclic alkene could have come from the corresponding ketone reacting with a Wittig reagent. The ketone precursor appears to have a methyl on the beta carbon of what was once an α,β-unsaturated ketone. That methyl could have been from a Gilman reagent, and the α,β-unsaturated ketone could easily have come from the oxidation of the secondary alcohol given as the starting material. Other solutions may be possible, but this one uses exclusively reactions mentioned in this chapter.

Chapter Test

True/False

1. 3-Methyl-2-butanone reacts with a given nucleophile faster than 3-methylbutanal does.

2. An S_N2 reaction by Ph_3P on R—X followed by butyl lithium forms a Wittig reagent.

Multiple Choice

3. Which reagents and/or conditions will convert 1-phenyl-3-pentanone to pentylbenzene?
   a. H_2, Pd, ethanol
   b. $NaBH_4$, then H_3O^+
   c. $Zn(Hg)$, H_3O^+
   d. All of the above
   e. None of the above

4. Which compound below would react fastest with a given nucleophile?
   a. cyclohexanone
   b. 2-methylcyclohexanone
   c. 2,2-dimethylcyclohexanone
   d. 2,2,6-trimethylcyclohexanone
   e. 2,2,6,6-tetramethylcyclohexanone

Short Answer

Provide structures for the following:

   5. Trichloroethanal

   6. Benzophenone

Name the following:

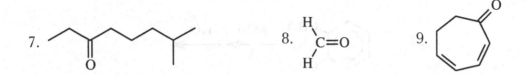

   10. Complete the following reaction with the structure of the organic product.

$$CH_3CH_2\overset{\displaystyle O}{\overset{\displaystyle \|}{C}}{-}Cl \quad + \quad [(\langle\bigcirc\rangle{-}CH_2)_2Cu]Li \quad \longrightarrow$$

   11. Show how one could make 2-hexanone from 1-hexyne.

   12. Show how one could make hexanal from 1-hexyne.

   13. Show how one could make *p*-methoxyacetophenone by a Friedel-Crafts acylation.

   14. Show the product that results from treating 5-hydroxypentanal with Jones' reagent.

   15. Show the product that results from treating 5-hydroxypentanal with Tollens' reagent.

   16. Show the structure of the product that results when 5-hydroxypentanal undergoes an intramolecular hemiacetal formation.

Provide a structural formula for the product that results from the reaction of cyclopentanone with each of the following:

   17. $CH_3CH_2CH_2MgBr$, then H_3O^+

   18. $NaBH_4$, then H_3O^+

   19. HCN

   20. $PhNHNH_2$ (phenylhydrazine)

   21. $CH_3CH_2CH_2NH_2$

   22. $(CH_3CH_2CH_2)_2NH$

   23. $HONH_2$

   24. $Ph_3P{=}CHCH_2CH_3$

   25. KOH, NH_2NH_2, heat

   26. One equivalent of $CH_3CH_2CH_2OH$ with mild acid catalyst

   27. Two equivalents of $CH_3CH_2CH_2OH$ with mild acid catalyst

Show the products that result from reaction of 3-methyl-2-cyclohexenone with each of the following:

28. O_3, then Zn and H_3O^+

29. $(CH_3CH_2CH_2)_2NH$

30. $CH_3CH_2CH_2MgBr$, then H_3O^+

31. $(CH_3CH_2CH_2)_2CuI$, then H_3O^+

32. Show the product of the following intramolecular Cannizzaro reaction.

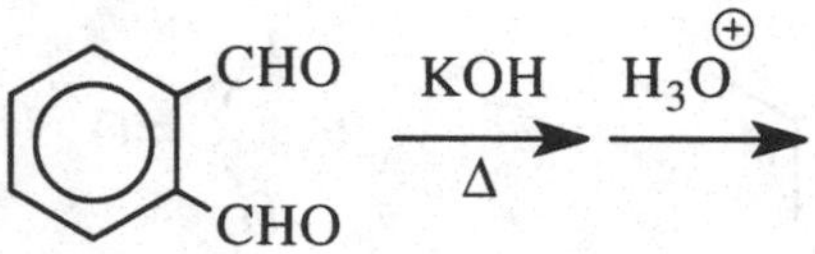

Chapter Test: Answers

1. **False**
2. **True**
3. **c.** Zn(Hg), H_3O^+
4. **a.** cyclohexanone
5. $Cl_3C{-}CH{=}O$
6. $Ph_2C{=}O$
7. 7-Methyl-3-octanone
8. Methanal or formaldehyde
9. 2,4-Cycloheptadienone

10. 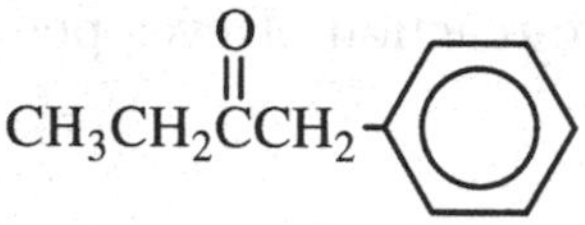

11. 12. 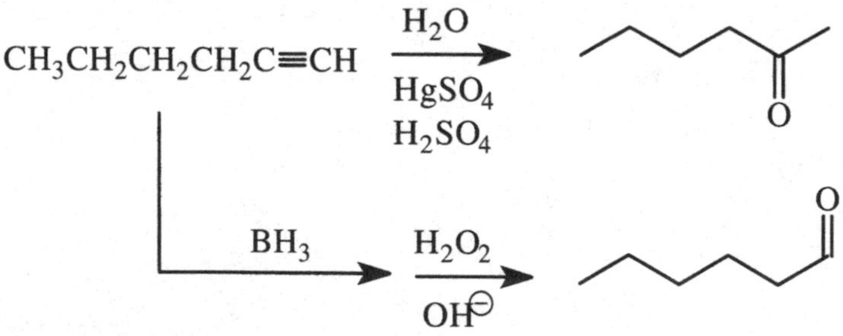

13. 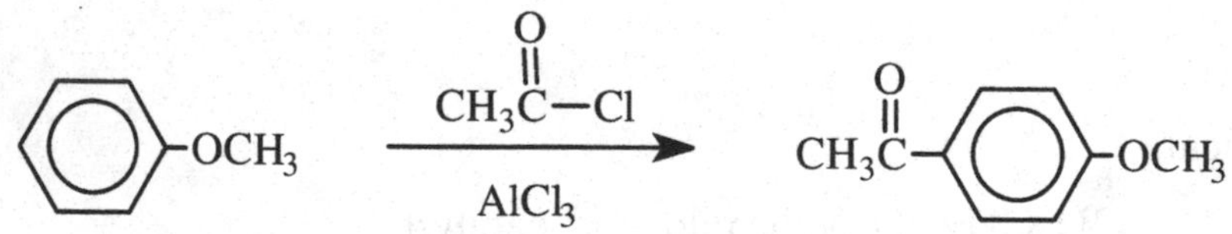

14. $HO\overset{O}{\overset{\|}{C}}(CH_2)_3\overset{O}{\overset{\|}{C}}OH$ 15. $HOCH_2CH_2CH_2CH_2\overset{O}{\overset{\|}{C}}OH$ 16. 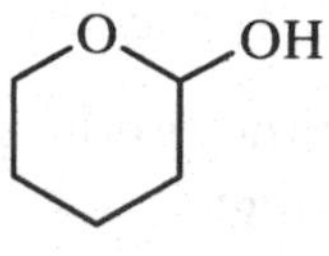

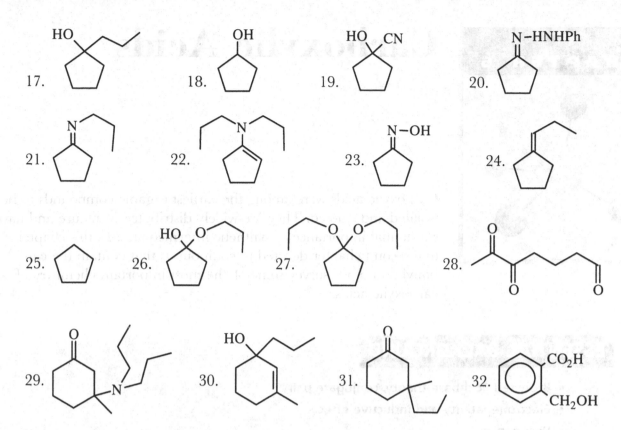

17.   18.   19.   20.

21.   22.   23.   24.

25.   26.   27.   28.

29.   30.   31.   32.

Check Your Performance

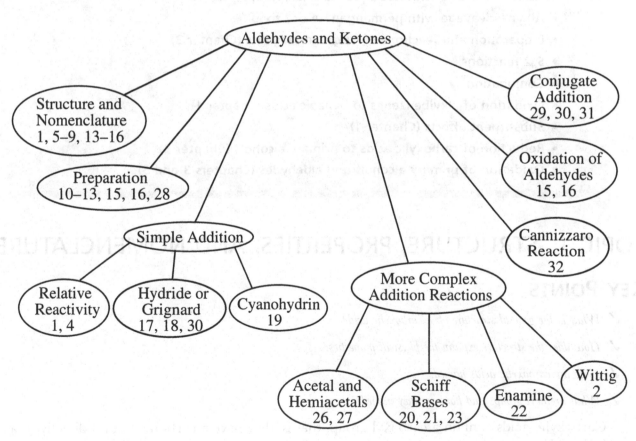

Note the number of questions in each grouping that you got wrong on the chapter test. Identify areas where you need further review and go back to relevant parts of this chapter.

CHAPTER 6

Carboxylic Acids

Carboxylic acids were among the earliest organic compounds to be isolated and studied. They are widely distributed in nature and have substantial importance as synthetic intermediates. In this chapter (the second chapter devoted to compounds that contain the carbonyl group) we survey some of the most important chemistry of carboxylic acids.

ESSENTIAL BACKGROUND

- Bronsted acid-base theory, conjugate pairs
- Electronegativity and inductive effects
- Resonance
- Alkene cleavage with hot aqueous acidic $KMnO_4$
- Alkyne cleavage with permanganate or ozone
- Preparation and reactions of Grignard reagents (Chapter 3)
- S_N2 reactions
- Conjugation
- Oxidation of alkylbenzenes to benzoic acids (Chapter 1)
- Substituent effects (Chapter 1)
- Reduction of carboxylic acids to primary alcohols (Chapter 3)
- Oxidation of primary alcohols and aldehydes (Chapters 3 and 5)

TOPIC 1: STRUCTURE, PROPERTIES, AND NOMENCLATURE

KEY POINTS

✓ *What is the general structure of a carboxylic acid?*

✓ *How does the structure explain the physical properties?*

✓ *How are carboxylic acids named?*

✓ *What are acyl groups and how are they named?*

Carboxylic acids contain a carboxyl group, that is, they have a carbonyl bound directly to a hydroxyl group. There are several common ways to represent these, which reveal different levels of structural detail.

$$\underset{R}{\overset{O}{\underset{|}{\overset{||}{C}}}}\underset{O-H}{} \quad = \quad \underset{R}{\overset{O}{\underset{|}{\overset{||}{C}}}}\underset{OH}{} \quad = \quad RCOOH \quad = \quad RCO_2H$$

As one would predict from the O—H and the other polar bonds that make up the carboxyl group, carboxylic acids have high melting and boiling points relative to other kinds of organic compounds. They are hydrogen bond donors and acceptors and participate in hydrogen bonding with themselves and polar organic solvents. Most small carboxylic acids are at least partially water soluble. Many carboxylic acids have strong, sharp, and often unpleasant odors. The smells of vinegar or aged perspiration are due to carboxylic acids.

Both common and systematic IUPAC names for carboxylic acids are similar to the names of the analogous aldehydes (Chapter 5). Simple carboxylic acids are named by finding the longest chain containing the carboxyl and replacing the "e" suffix of the corresponding alkane parent with the "oic acid." The carbonyl carbon is defined as C1 and substituents are included as prefixes with numbers as locators. If R is cyclic, the name is derived from the corresponding parent molecule with the suffix "carboxylic acid." There are also some widely used common names you should know. These nomenclatures are illustrated in **Table 6.1**.

Dicarboxylic acids of the form $HO_2C(CH_2)_nCO_2H$ are named by adding "dioic acid" as a suffix onto the parent alkane having the same number of carbons. These dicarboxylic acids have recognized common names used mostly by biochemists. These names are not obvious or systematic in most cases, but the mnemonic "Oh my! Such good apples" will give you the first letter of the common names in order of increasing dicarboxylic acid size. These are shown in **Table 6.2**.

Table 6.1 Structures and Names of Some Representative Carboxylic Acids

STRUCTURE	SYSTEMATIC IUPAC	COMMON
HCO_2H	Methanoic acid	Formic acid
CH_3CO_2H	Ethanoic acid	Acetic acid
$CH_3CH_2CO_2H$	Propoanoic acid	Propionic acid
$CH_3(CH_2)_2CO_2H$	Butanoic acid	Butyric acid
$CH_3(CH_2)_3CO_2H$	Pentanoic acid	Valeric acid
⬡—CO_2H	Benzenecarboxylic acid	Benzoic acid
—CO_2H	2-Ethylbutanoic acid	
$CH_3CHCHCO_2H$ (OH, CH_3)	3-Hydroxy-2-methylbutanoic acid	
—CO_2H	Cyclohexanecarboxylic acid	
—CO_2H	1-Cyclopentenecarboxylic acid	

Table 6.2 Names and Structures of Some Dicarboxylic Acids

STRUCTURE	IUPAC NAME	COMMON NAME
HO_2CCO_2H	Ethanedioic acid	Oxalic acid
$HO_2CCH_2CO_2H$	Propanedioic acid	Malonic acid
$HO_2C(CH_2)_2CO_2H$	Butanedioic acid	Succinic acid
$HO_2C(CH_2)_3CO_2H$	Pentanedioic acid	Glutaric acid
$HO_2C(CH_2)_4CO_2H$	Hexanedioic acid	Adipic acid

It is often desirable to refer to part of the structure of a carboxylic acid. A group that looks like a carboxylic acid without its OH is called an **acyl group**. (Recall placing these onto aromatic rings in the Friedel-Crafts acylation of Chapter 1.) These are named by replacing the "ic acid" of the carboxylic acid name with "yl." Some representative acyl groups and names are shown below.

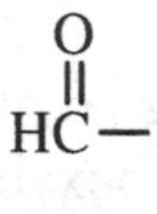

| Formyl | 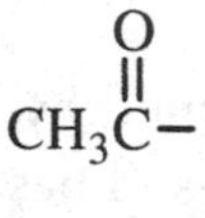 Acetyl | 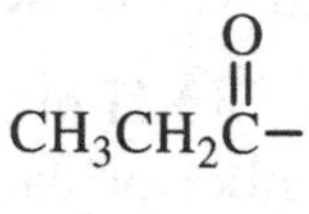 Propanoyl | 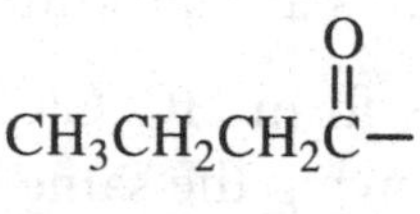 Butanoyl | 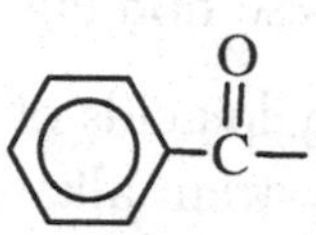 Benzoyl |

Topic Test 1: Structure, Properties, and Nomenclature

True/False

1. Carboxylic acid names usually end in "yl."

2. Decanoic acid has the structure $CH_3(CH_2)_{10}CO_2H$.

Multiple Choice

3. The group below is

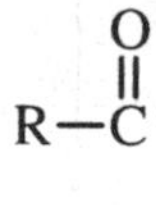

   a. a carboxyl group.
   b. an alkyl group.
   c. a carbonyl.
   d. an acyl group.
   e. None of the above

4. Which compound below has the highest boiling point?
   a. 1-Hexanol
   b. Dipropyl ether
   c. Hexane
   d. Hexanoic acid
   e. Hexanedioic acid

Short Answer

5. Provide an unambiguous structural formula for (*E*)-3-methyl-2-octenoic acid.

6. Name the following compound.

Topic Test 1: Answers

1. **False.** Carboxylic acid names usually end in "ic acid" or "oic acid."

2. **False.** Decanoic acid has 10 carbons and has the structure $CH_3(CH_2)_8CO_2H$. The 12-carbon structure shown would be called dodecanoic acid.

3. **d.** An acyl group. Answers a through c go with the structures CO_2H, R and C=O, respectively.

4. **e.** Hexanedioic acid. Besides having the highest molecular mass of the choices given, hexanedioic acid has two carboxyl groups that thus has the greatest amount of intermolecular attraction through hydrogen bonding, giving the compound a high boiling point.

5.

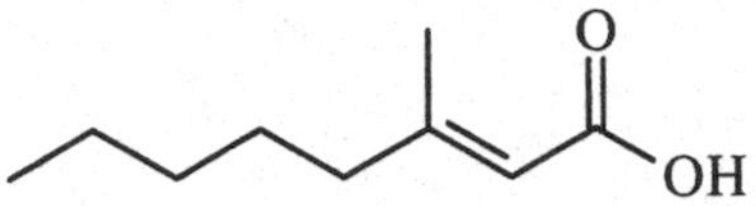

6. *trans*-1,3-Cyclobutanedicarboxylic acid

TOPIC 2: PREPARATION OF CARBOXYLIC ACIDS

KEY POINTS

✓ *What reactions from previous chapters yield carboxylic acids?*

✓ *How are Grignard reagents converted to carboxylic acids?*

✓ *Under what circumstances is RX converted to RCO_2H via a Grignard reagent?*

✓ *How are nitriles converted to carboxylic acids?*

✓ *Under what circumstances is RX converted to RCO_2H via a nitrile?*

Cleavage of alkenes and alkynes yields carbonyl compounds. Alkenes bearing vinylic hydrogens are cleaved with acidic or basic aqueous potassium permanganate to aldehydes that oxidize under the reaction conditions to yield carboxylic acids. Permanganate or ozone with aqueous acid cleaves alkynes to carboxylic acids.

$$RCH=CH_2 \xrightarrow[\text{H}_3\text{O}^\oplus \text{ or } \text{OH}^\ominus]{\text{KMnO}_4} \left[\begin{array}{c} R \\ C=O \\ H \end{array} \quad O=C\begin{array}{c} H \\ H \end{array} \right] \longrightarrow \left[O=C\begin{array}{c} OH \\ OH \end{array} \right]$$

$$MnO_2 + \boxed{\begin{array}{c} R \\ C=O \\ OH \end{array}} + \boxed{CO_2 + H_2O}$$

$$R-C\equiv C-R' \xrightarrow[\substack{\text{or} \\ O_3, \text{H}_3\text{O}^\oplus}]{\substack{\text{KMnO}_4 \\ \text{H}_3\text{O}^\oplus}} RCO_2H + HO_2CR'$$

$$R-C\equiv C-H \xrightarrow[\substack{\text{or} \\ O_3, \text{H}_3\text{O}^\oplus}]{\substack{\text{KMnO}_4 \\ \text{H}_3\text{O}^\oplus}} RCO_2H + \left[\begin{array}{c} O \\ \| \\ HOCOH \end{array} \right] \longrightarrow \begin{array}{c} CO_2 \\ + \\ H_2O \end{array}$$

Primary alcohols and aldehydes are oxidized by a variety of reagents to yield carboxylic acids (Chapters 3 and 5).

$$\begin{array}{c} RCH_2OH \\ \text{or} \\ RCHO \end{array} \xrightarrow[\text{KMnO}_4, \text{H}_3\text{O}^\oplus]{\substack{\text{Jones' reagent or} \\ K_2Cr_2O_7, \text{H}_3\text{O}^\oplus \text{ or}}} RCO_2H$$

$$RCHO \xrightarrow[\text{Ag(NH}_3)_2^\oplus]{\text{Tollens' reagent}} \xrightarrow{\text{H}_3\text{O}^\oplus}$$

When alkylbenzenes bearing benzylic hydrogen atoms are treated with hot aqueous permanganate, all the carbons accept the benzylic carbon are cleaved off and the resulting product is a benzoic acid derivative as in the examples below (Chapter 1).

There are two methods for replacing the halogen of an alkyl halide with a CO_2H group. The alkyl halide can be converted into a Grignard reagent and then reacted with carbon dioxide. This strategy is limited to those organic halides that are compatible with formation of Grignard reagents (i.e., no protic groups like OH or NH in the solvent or starting material) but has the advantage of working for vinylic and aromatic halides and simple alkyl halides.

$$R-X \xrightarrow[\text{Ether}]{\text{Mg}} RMgX \xrightarrow{CO_2} \xrightarrow{H_3O^+} R-CO_2H$$

An alternative method uses cyanide ion as a nucleophile to replace the halogen to yield a nitrile that can be hydrolyzed with acid or base catalyst to produce the carboxylic acid.

$$R-X \xrightarrow[\text{S}_N2]{CN^-} R-C\equiv N \xrightarrow[\text{H}_3O^+ \text{ or } \overline{O}H]{H_2O} R-CO_2H$$

Because the first step is an S_N2 reaction, this strategy will not work for tertiary, vinylic, or aromatic halides but is generally successful for simple methyl and primary R—X.

Topic Test 2: Preparation of Carboxylic Acids

True/False

1. Aldehydes will oxidize to carboxylic acids if treated with acidic permanganate.

2. A primary alcohol will oxidize to a carboxylic acid if treated with Tollens' reagent.

Multiple Choice

3. Which reaction below will yield acetic acid?
   a. Treating 2-butene with acidic aqueous potassium permanganate.
   b. Treating 2-butyne with ozone and aqueous acid.
   c. Treating ethanal with Tollens' reagent then aqueous acid.
   d. All of the above
   e. None of the above

4. What reactant below would yield benzoic acid when treated with hot aqueous potassium permanganate?
   a. *p*-Xylene (1,4-dimethylbenzene)
   b. $PhCH_2CH_2Ph$
   c. Benzene
   d. All of the above
   e. None of the above

Short Answer

Show the reagents and/or conditions one could use to perform the following transformations. More than one step may be required.

5. $HOCH_2CH_2CH_2Br \rightarrow HOCH_2CH_2CH_2CO_2H$

6. Bromobenzene $\rightarrow$ benzoic acid

Topic Test 2: Answers

1. **True.** This reaction is a way to make RCO_2H directly from the analogous aldehyde, and it is also the reason that any aldehydes initially produced by permanganate cleavage of alkenes will be further oxidized under the reaction conditions to yield RCO_2H.

2. **False.** Although several oxidizing agents convert RCH_2OH to RCO_2H, Tollens' reagent is not among them. Instead, one could use acidic permanganate, Jones' reagent, or potassium dichromate in aqueous acid. (Tollens' reagent is selective for oxidizing only aldehydes to $RCOO^-$ that in turn are protonated upon treatment with acid to give RCOOH.)

3. **d.** All of the above

4. **b.** $PhCH_2CH_2Ph$. This reactant yields two equivalents of benzoic acid upon treatment with hot aqueous permanganate. (Think of it as $PhCH_2R$ on both sides.) The reactant from choice a would lead to 1,4-benzenedicarboxylic acid. The reactant from choice c would likely not react at all but in any case could not possibly produce a seven-carbon carboxylic acid from oxidation of a six-carbon reactant.

5. KCN (or some other cyanide salt) in a polar aprotic solvent gives an S_N2 reaction yielding $HOCH_2CH_2CH_2CN$, which is subsequently treated with water and acid or base catalyst to give the desired product. Note that one should *not* attempt to use the Grignard then CO_2 method in this case because the hydroxyl in the reactant will prevent successful formation of a Grignard reagent.

6. Treat bromobenzene with Mg in dry ether, then CO_2, then H_3O^+. Note that one must *not* attempt to use the cyanide method for this problem because the aryl halide starting material is not capable of the required S_N2 reaction and thus will not yield PhCN in the first step of that method.

TOPIC 3: ACID REACTIONS OF CARBOXYLIC ACIDS

KEY POINTS

✓ *How acidic are carboxylic acids relative to other acids and other organic compounds?*

✓ *What are the products of an acid-base reaction involving a carboxylic acid?*

✓ *How are the conjugate bases of carboxylic acids named?*

✓ *How are K_a and pK_a defined and how are they interpreted?*

✓ *What effect will a given substituent have on the acid strength of a carboxylic acid?*

As the name implies, carboxylic acids are acidic. They are generally more acidic than other classes of organic compounds discussed, such as alcohols and phenols, but are still considered weak acids in that they do not normally ionize completely in water.

The conjugate base of a carboxylic acid is called a **carboxylate**. Carboxylic acids react with bases to yield carboxylate salts.

$$RCO_2H \quad + NaOH_{(aq)} \rightarrow \quad [RCO_2^-]Na^+_{(aq)} \quad + H_2O$$
$$\text{Carboxylic acid} \qquad\qquad \text{Sodium carboxylate}$$

The name of a carboxylate anion is easily derived by replacing the "ic acid" or "oic acid" with the suffix "ate." This is similar to the way ions from inorganic oxygen-containing acids are named (e.g., nitric acid → nitrate).

$$HCO_2^{\ominus} \qquad CH_3CO_2^{\ominus} \qquad (CH_3)_2CHCO_2^{\ominus}$$

Formate   Acetate   2-Methylpropanoate
(Methanoate)   (Ethanoate)

Cyclobutanecarboxylate   Benzoate

The acidity of an acid, HA, is quantified and reported numerically as K_a or pK_a. These are defined as

$$K_a = \frac{[H_3O^+][A^-]}{[HA]} \quad \text{and} \quad pK_a = -\log K_a$$

The stronger the acid, the more it will ionize in water, and the numerator of the K_a expression will increase while the denominator decreases. Therefore, a stronger acid will have a larger K_a. Because a negative log is involved, a stronger acid will have a smaller pK_a (i.e., less positive or more negative). Some acids are listed in **Table 6.3** along with their values of K_a and pK_a.

Substituents can affect the strength of a carboxylic acid. In general, electron-withdrawing groups enhance acidity, whereas electron-donating groups inhibit acidity. The electron withdrawal or donation can be inductive (electronegativity) or via resonance.

← Increasing acid strength

$$EWG-CH_2CO_2H \quad > \quad CH_3CO_2H \quad > \quad EDG-CH_2CO_2H$$

CO$_2$H   CO$_2$H   CO$_2$H

EWG   >   >   EDG

Table 6.3 Some K_a and pK_a Values for Common Compounds			
STRUCTURE AND NAME		K_a	pK_a
HCl	Hydrochloric acid	~10^7	~-7
HNO$_3$	Nitric acid	2.5×10^1	-1.4
HCO$_2$H	Formic acid	1.8×10^{-4}	3.75
PhCO$_2$H	Benzoic acid	6.5×10^{-5}	4.19
CH$_3$CO$_2$H	Acetic acid	1.8×10^{-5}	4.75
PhOH	Phenol	1.3×10^{-10}	9.89
CH$_3$CH$_2$OH	Ethanol	~10^{-16}	~16

Topic Test 3: Acid Reactions of Carboxylic Acids

True/False

1. Formate ion is the major organic species present in an aqueous solution of formic acid.

2. An acid with a pK_a of 3 is about 100 times stronger than an acid with a pK_a of 5.

Multiple Choice

3. The conjugate base of a carboxylic acid is
   a. hydroxide.
   b. water.
   c. a carboxylate ion.
   d. an inorganic salt.
   e. None of the above

4. The name of the organic product that results from reaction of propanoic acid with potassium hydroxide is
   a. 2-hydroxypropanoic acid.
   b. potassium propanide.
   c. propanoic potassiate.
   d. potassium propanoate.
   e. None of the above

Short Answer

Order the following according to acidity:

5. $CH_3CH_2CH_2CO_2H$, $CH_3OCH_2CO_2H$, CH_3CO_2H

6. Benzoic acid, *p*-methoxybenzoic acid, *p*-cyanobenzoic acid

Topic Test 3: Answers

1. **False.** Although formic acid does partially ionize in water (to give hydronium ion and formate ion), the equilibrium favors the un-ionized acid HCOOH. (Formic acid is a weak acid and therefore reacts only slightly in water.)

2. **True.** Because pK_a is a logarithmic quantifier, each pK_a unit represents an order of magnitude in acid strength. Recall also that a smaller pK_a (more negative or less positive) corresponds to a stronger acid.

3. **c.** a carboxylate ion.

4. **d.** potassium propanoate.

5. Note that each of these has the form G—CH_2CO_2H, where G is ethyl, methoxy, or hydrogen, respectively. When G is methoxy, the electronegative oxygen inductively withdraws electron density relative to what is expected for G = H. Similarly, compare the mild electron-donating ability of CH_3CH_2 relative to H. This predicts the order of acidity is $CH_3OCH_2CO_2H > CH_3CO_2H > CH_3CH_2CH_2CO_2H$.

6. As in the previous problem, the key here is to recognize that these compounds differ only by the atom or group occupying the *para* position. H, CH_3O, and $N\equiv C$, respectively. The methoxy is an electron-donor group in this case because the lone pair of electrons on oxygen is conjugated with the aromatic ring, resulting in decreased acidity relative to benzoic acid. The unsaturation of the cyano group makes it an electron-withdrawing group that will enhance the acidity, making the predicted order of acidity *p*-cyanobenzoic acid > benzoic acid > *p*-methoxybenzoic acid.

TOPIC 4: OTHER REACTIONS OF CARBOXYLIC ACIDS

KEY POINTS

✓ *What reagents reduce carboxylic acids?*

✓ *What products result from carboxylic acid reduction?*

✓ *What is the general form of an alpha substitution reaction?*

✓ *What is the general form of a nucleophilic acyl substitution?*

✓ *How are carboxylic acids converted into acyl halides?*

Some reactions of carboxylic acids are not acid-base reactions. Recall from Chapter 3 that carboxylic acids can be reduced to primary alcohols with lithium aluminum hydride followed by aqueous acid. Alternatively, borane can be used as the reducing agent.

$$R-\overset{\overset{\displaystyle O}{\|}}{C}-OH \xrightarrow[\text{THF}]{\text{LiAlH}_4 \text{ or } \text{BH}_3} \xrightarrow{H_3O^{\oplus}} R-CH_2OH$$

The position alpha to the carbonyl can sometimes be the site of a carboxylic acid's reactivity. When some other atom or group, "E," replaces alpha hydrogen, the process is called **alpha substitution**. We will examine such reactions more in Chapter 8 but for now you should recognize the general form of an alpha substitution.

$$-\overset{|}{\underset{H}{C}}-\overset{\overset{\displaystyle O}{\|}}{C}-OH \xrightarrow{\text{"E}^+\text{"}} -\overset{|}{\underset{E}{C}}-\overset{\overset{\displaystyle O}{\|}}{C}-OH$$

Some reactions of carboxylic acids result in replacement of the hydroxyl with another atom or group. A reaction of this type is called a **nucleophilic acyl substitution**. These will be examined in Chapter 7 but for now you should recognize the general form of such a process and be familiar with the examples where carboxylic acids are converted to acyl chlorides or bromides with thionyl chloride or phosphorus tribromide, respectively.

$$R-\overset{\overset{\displaystyle O}{\|}}{C}-OH \xrightarrow{\text{"Nu}^-\text{"}} R-\overset{\overset{\displaystyle O}{\|}}{C}-Nu$$

$$R-\overset{\overset{\displaystyle O}{\|}}{C}-OH \xrightarrow{\text{SOCl}_2} R-\overset{\overset{\displaystyle O}{\|}}{C}-Cl$$

$$R-\overset{\overset{\displaystyle O}{\|}}{C}-OH \xrightarrow{\text{PBr}_3} R-\overset{\overset{\displaystyle O}{\|}}{C}-Br$$

Topic Test 4: Other Reactions of Carboxylic Acids

True/False

1. Carboxylic acids can be converted to acyl chlorides with thionyl chloride in a process called nucleophilic acyl substitution.

2. Lithium aluminum hydride will reduce a carboxylic acid to an aldehyde.

Multiple Choice

3. The reaction below is an example of what kind of transformation?

$$CH_3CH_2C\overset{O}{\underset{OH}{\diagdown}} \xrightarrow[\text{heat}]{NH_2CH_3} CH_3CH_2C\overset{O}{\underset{NHCH_3}{\diagdown}}$$

    a. Alpha substitution
    b. Reduction
    c. Nucleophilic acyl substitution
    d. All of the above
    e. None of the above

4. The reaction below is an example of what kind of transformation?

$$CH_3CH_2C\overset{O}{\underset{OH}{\diagdown}} \longrightarrow CH_3CHC\overset{O}{\underset{OH}{\diagdown}}, \underset{Br}{|}$$

    a. Alpha substitution
    b. Reduction
    c. Nucleophilic acyl substitution
    d. All of the above
    e. None of the above

Short Answer

Provide unambiguous structural formulas for the missing organic compounds.

5. (bicyclic)—CO$_2$H $\xrightarrow[\text{THF}]{BH_3}$ $\xrightarrow{H_3O^{\oplus}}$

6. Br—(benzene ring)—CO$_2$H $\xrightarrow{SOCl_2}$

Topic Test 4: Answers

1. **True.** The OH of the acid is replaced by Cl in the process.

2. **False.** LiAlH$_4$ followed by H$_3$O$^+$ reduces a carboxylic acid to a primary alcohol.

3. **c.** Nucleophilic acyl substitution

4. **a.** Alpha substitution

5.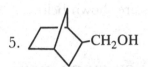

6.

APPLICATION

Long-chain carboxylic acids are often referred to as **fatty acids**. These are obtained from the hydrolysis of triacylglycerols: esters of fatty acids that are found in animal **fats** and vegetable **oils**. Fatty acids generally contain an even number of carbons because they are biosynthesized from two-carbon building blocks. The alkyl portions of these fatty acids can be **saturated** or can contain one or more alkene linkages (**monounsaturated** or **polyunsaturated**). The alkene bonds of most naturally occurring fatty acids have the *cis* configuration. Some common fatty acids are shown in **Table 6.4**.

DEMONSTRATION PROBLEM

Show how one could convert butanoic acid into pentanoic acid. More than one step will be required.

Solution

Building a five-carbon carboxylic acid from one with four carbons will obviously require adding a carbon at some point. As we saw in Topic 2, there are two ways to extend a chain by one carbon to yield a carboxylic acid, and both strategies start with an alkyl halide.

Table 6.4 Some Common Fatty Acids and Sources		
NAME	**STRUCTURE**	**SOURCE**
Saturated		
Lauric	CH$_3$(CH$_2$)$_{10}$CO$_2$H	Coconut oil
Myristic	CH$_3$(CH$_2$)$_{12}$CO$_2$H	Butter
Palmitic	CH$_3$(CH$_2$)$_{14}$CO$_2$H	Most fats and oils
Stearic	CH$_3$(CH$_2$)$_{16}$CO$_2$H	Most fats and oils
Unsaturated (all alkenes are cis)		
Oleic	CH$_3$(CH$_2$)$_7$CH=CH(CH$_2$)$_7$CO$_2$H	Olive oil
Linoleic	CH$_3$(CH$_2$)$_3$(CH$_2$CH=CH)$_2$(CH$_2$)$_7$CO$_2$H	Most vegetable oils
Linolenic	CH$_3$(CH$_2$CH=CH)$_3$(CH$_2$)$_7$CO$_2$H	Soybean oil

$$R\text{—}X \xrightarrow[\text{Ether}]{\text{Mg}} RMgX \xrightarrow{CO_2} \xrightarrow{H_3O^+} R\text{—}CO_2H$$

$$R\text{—}X \xrightarrow[S_N2]{CN^-} R\text{—}C\equiv N \xrightarrow[H_3O^+ \text{ or } OH^-]{H_2O} R\text{—}CO_2H$$

Either of these methods could be applied to 1-halobutane to give pentanoic acid so the problem is now simplified to finding a way to make a 1-halobutane from butanoic acid. Although we have not discussed any method for converting a carboxylic acid directly into an alkyl halide, the transformation could be carried out in several steps via the alcohol. There is more than one possible correct solution to this problem. Some possible reagent combinations are shown below.

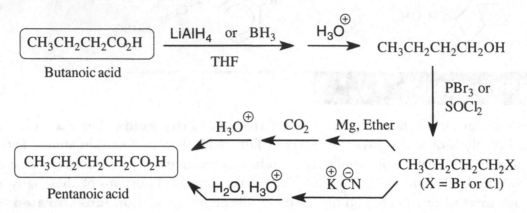

Chapter Test

True/False

1. The stronger an acid is, the larger its K_a will be.

2. Succinic acid is another name for butanedioic acid.

3. *p*-Nitrobenzoic acid has a smaller pK_a than does benzoic acid.

4. Carboxylic acids are less water soluble than ketones of similar size and shape.

5. The melting point of hexanoic acid is lower than that of 1-hexanol.

Multiple Choice

6. Which reagents below would convert 1-bromocyclopentene to 1-cyclopentenecarboxylic acid?
   a. $LiAlH_4$, then H_3O^+
   b. NaCN, then H_3O^+
   c. Mg, ether, then CO_2, then H_3O^+
   d. $KMnO_4$, H_3O^+
   e. None of the above

7. Which functional group below does Tollens' reagent selectively oxidize?
   a. Carboxylic acid
   b. Primary alcohol
   c. Aldehyde
   d. All of the above
   e. None of the above

8. Which acid below is the strongest?
   a. Butanoic acid
   b. 2-Chlorobutanoic acid
   c. 3-Chlorobutanoic acid
   d. 4-Chlorobutanoic acid
   e. All the above will have the same K_a.

9. Which acid below is the strongest?
   a. Benzoic acid
   b. *p*-Methylbenzoic acid
   c. *p*-(Trichloromethyl)benzoic acid
   d. *p*-(Trifluoromethyl)benzoic acid
   e. All the above will have the same K_a.

10. Which reaction is possible for pentanoic acid?
    a. Reaction with base to yield pentanoate salt
    b. Nucleophilic acyl substitution to yield an acyl halide
    c. Alpha substitution to yield pentanoic acid with a substituent on carbon number 2
    d. Reduction to 1-pentanol
    e. All of the above

11. The product of the reaction sequence shown below would be

$$CH_3CH_2C{\equiv}CCH_2CH_3 \xrightarrow[H_3O^\oplus]{KMnO_4} \xrightarrow{BH_3} \xrightarrow{H_3O^\oplus}$$

   a. *cis*-3-hexene.
   b. *trans*-3-hexene.
   c. propanoic acid.
   d. 1-propanol.
   e. None of the above

12. The group below would be called

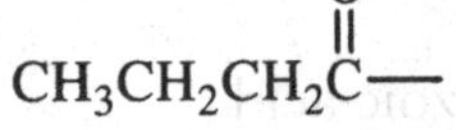

$$CH_3CH_2CH_2\overset{\overset{\displaystyle O}{\|}}{C}{-}$$

   a. butanoyl.
   b. propanoyl.
   c. butanoate.
   d. propanoate.
   e. None of the above

Short Answer

Complete the following reaction schemes with unambiguous structural formulas for the missing organic compounds.

13.

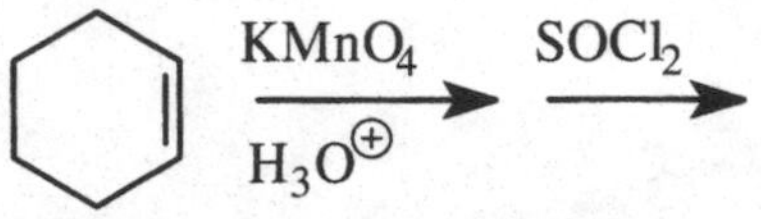

$$\xrightarrow[\text{H}_3\text{O}^\oplus]{\text{KMnO}_4} \xrightarrow{\text{SOCl}_2}$$

14.

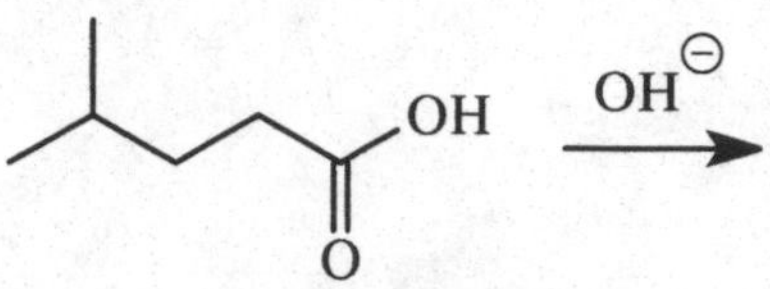

$$\xrightarrow{\text{OH}^\ominus}$$

15. ? (C_6H_{10}) $\xrightarrow[\text{H}_3\text{O}^\oplus]{\text{O}_3}$ $+$

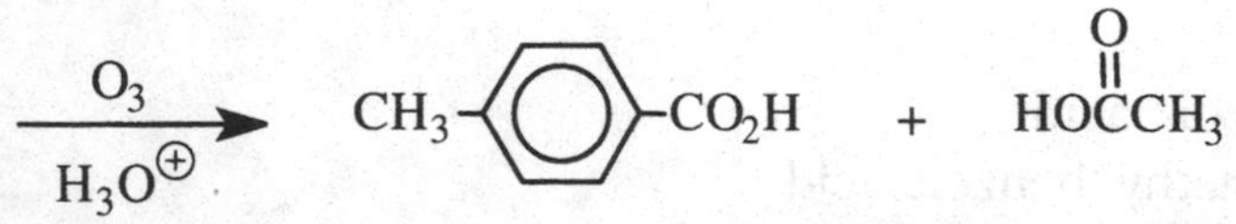

16. Provide a name and structure for the product that results from neutralization of $(CH_3)_2CHCH_2CO_2H$ with sodium hydroxide.

17. Provide a structure for the organic product that results from treating 2-methylbenzaldehyde with Tollens' reagent, then aqueous acid, then PBr_3.
Name the following

18. $(CH_3)_2CHCH_2CH_2CH_2CH_2CO_2H$

19. $(CH_3CO_2)_2$ Mg

20. What is the structure of a formyl group?

Chapter Test: Answers

1. **True**

2. **True**

3. **True**

4. **False**

5. **False**

6. **c.** Mg, ether, then CO_2, then H_3O^+

7. **c.** Aldehyde

8. **b.** 2-Chlorobutanoic acid

9. **d.** p-(Trifluoromethyl)benzoic acid

10. **e.** All of the above

11. **d.** 1-Propanol

12. **a.** Butanoyl

13. 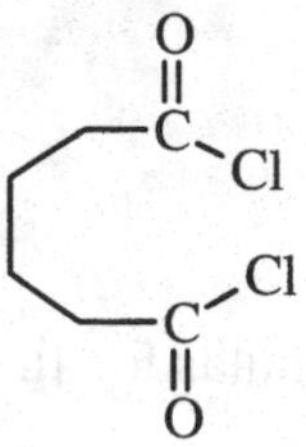   14. 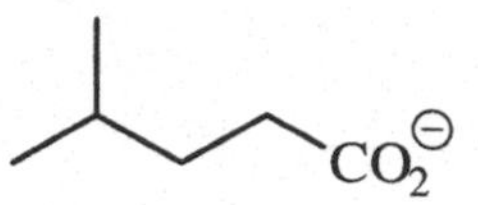   15. 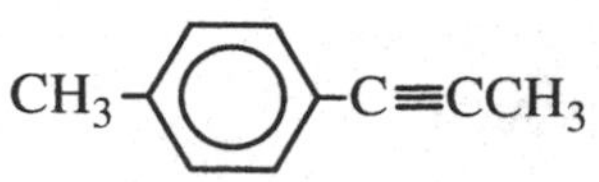 CH$_3$——C≡CCH$_3$

16. $(CH_3)_2CHCH_2CO_2^-Na^+$, sodium 3-methylbutanoate

17.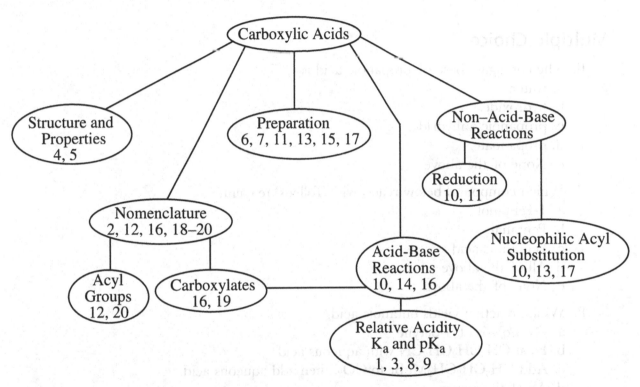

18. 6-Methylheptanoic acid

19. Magnesium acetate (or magnesium ethanoate)

20. H—C=O

Check Your Performance

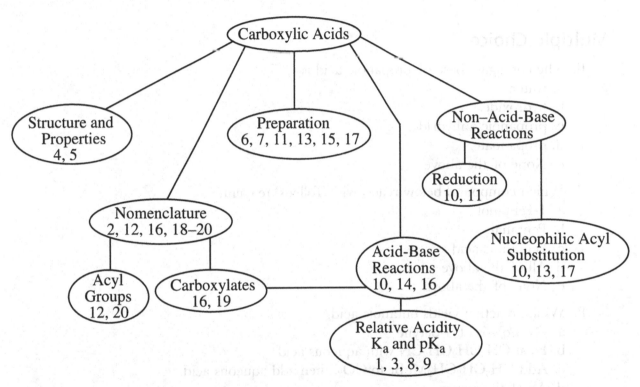

Note the number of questions in each grouping that you got wrong on the chapter test. Identify areas where you need further review and go back to relevant parts of this chapter.

Midterm Exam

True/False

1. Benzoic acid will nitrate faster than toluene when treated with a mixture of sulfuric and nitric acids.

2. The group SO_3H is a deactivating meta-director for EAS reactions.

3. Nitrobenzene does not usually react under Friedel-Crafts conditions.

4. The ^1H nuclear magnetic resonance signal pattern produced by a *tert*-butyl group appears as three singlets that are the same size.

5. Tertiary alcohols are not oxidized by Jones' reagent.

Multiple Choice

6. The conjugate base of propanoic acid is
   a. water.
   b. hydroxide.
   c. propanoic anhydride.
   d. propanoate.
   e. None of the above

7. Which compound below reacts with Tollens' reagent?
   a. 1-Pentanol
   b. Pentanal
   c. Pentanoic acid
   d. All of the above
   e. None of the above

8. Which reaction yields butanoic acid?
   a. Ozonolysis of 4-octyne
   b. Heat $CH_3CH_2CH_2CN$ with aqueous acid
   c. Add $CH_3CH_2CH_2MgBr$ to CO_2, then add aqueous acid
   d. All of the above
   e. None of the above

9. The group COOH is called a(n)
   a. carbonyl.
   b. carboxyl.
   c. acyl group.
   d. carbonyl oxide.
   e. None of the above

10. Which of the following is most acidic?
    a. Phenol
    b. Cyclohexanol
    c. Benzoic acid
    d. *p*-Ethylbenzoic acid
    e. All the above have approximately the same pK_a.

11. Which conditions below yield no reaction with toluene?
    a. *N*-bromosuccinimide
    b. Hot, acidic, aqueous potassium permanganate
    c. H_2, Pd
    d. Br_2, $FeBr_3$
    e. All of the above react with toluene.

12. Another name for a thiol is a
    a. hemiacetal.
    b. mercaptan.
    c. disulfide.
    d. ketone.
    e. None of the above

13. Which of the following has the highest boiling point?
    a. $HOCH_2CH_2CH_2OH$
    b. $CH_3OCH_2OCH_3$
    c. $CH_3OCH_2CH_2OH$
    d. $HOCH_2CH_2CH_3$
    e. All the above have three carbons and therefore approximately the same boiling point.

14. Which of the following is most acidic?

    e. All the above the same acidity.

Specify mass spectroscopy, ultraviolet visible, infrared, or nuclear magnetic resonance according to which one is best associated with each item below.

15. Bond lengths and bond angles are fluctuating.

16. Shows the number of different carbon environments.

17. Shows the kinds of bonds present, i.e., can indicate functional groups.

18. Usually indicates the molecular weight.

19. Electrons are moved from a lower energy orbital to a higher energy orbital.

20. Molecules are bombarded by high-energy electrons resulting in ionization of the molecules that often break apart afterward.

21. Spinning nuclei in a magnetic field absorb radio frequencies.

22. Indicates the presence and extent of conjugation.

Short Answer

Provide unambiguous structural formulas for the compounds named below.

23. Benzyl alcohol

24. Tetrahydrofuran

25. Dibutyl sulfide

26. *m*-Mercaptobenzaldehyde

Name the following:

27. $CH_3CH_2CH{=}CH(CH_2)_5CO_2H$

28.

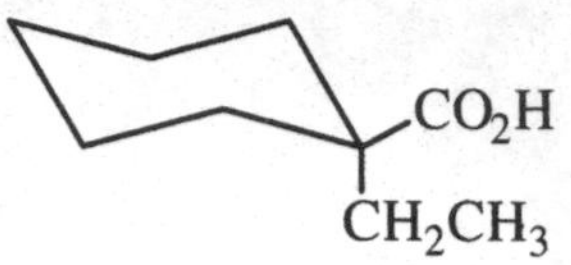

29. $\underset{\quad OH}{CH_3CH_2CH}(CH_2)_5\underset{\quad CH_3}{CHCH_3}$

30.

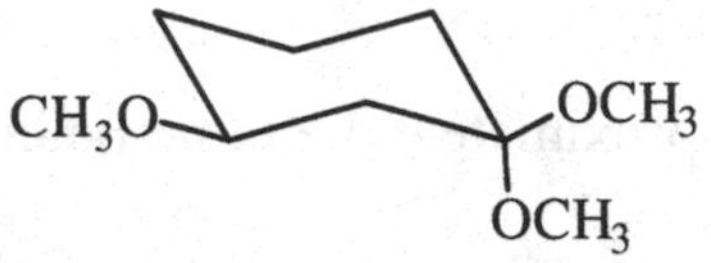

31.

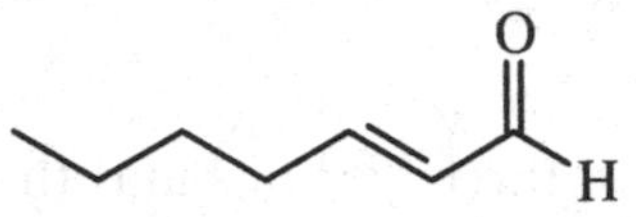

32. $(CH_3)_3CCH_2CH_2\overset{O}{\overset{\|}{C}}CH_2CH_2CH_3$

33.

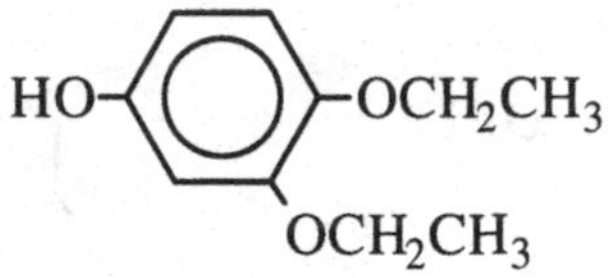

34.

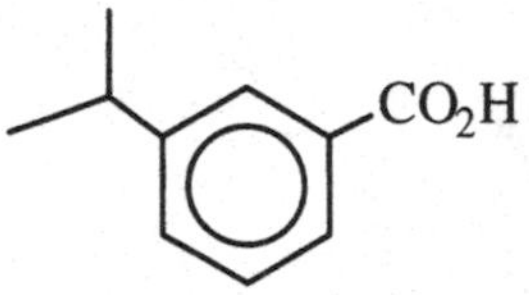

Provide unambiguous structural formulas for the missing organic compounds.

35. 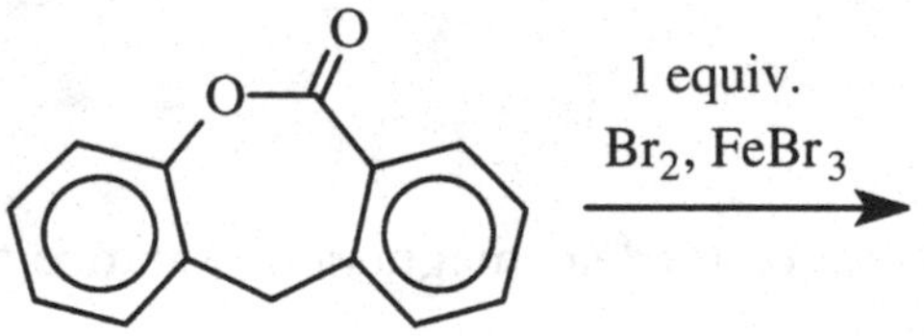 $\xrightarrow[\text{Br}_2,\ \text{FeBr}_3]{\text{1 equiv.}}$

36. 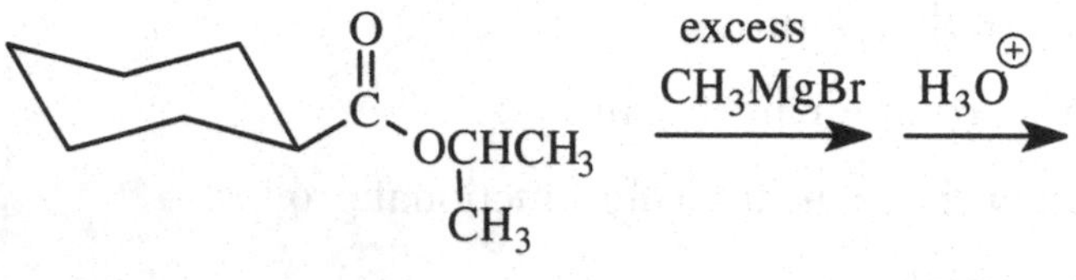 $\xrightarrow[\text{CH}_3\text{MgBr}]{\text{excess}}$ $\xrightarrow{\text{H}_3\text{O}^{\oplus}}$

37. 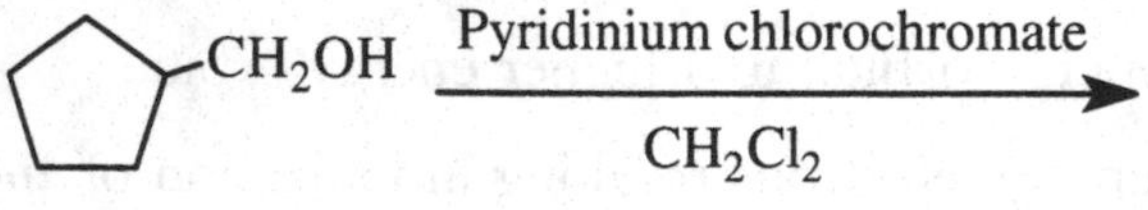 $\xrightarrow[\text{CH}_2\text{Cl}_2]{\text{Pyridinium chlorochromate}}$

38. 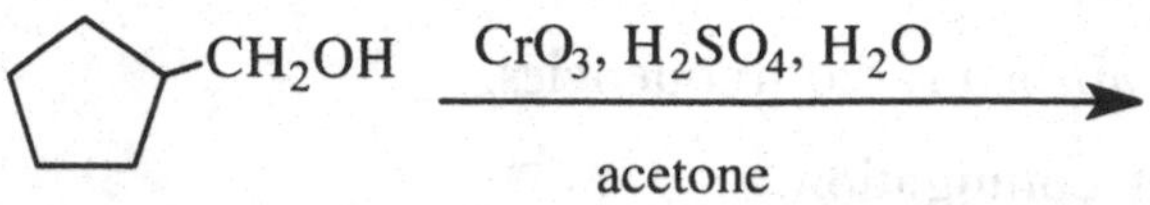 $\xrightarrow[\text{acetone}]{\text{CrO}_3,\ \text{H}_2\text{SO}_4,\ \text{H}_2\text{O}}$

39.

$\xrightarrow[\text{pyridine}]{\text{POCl}_3}$

40.

$\xrightarrow[\text{CH}_2\text{Cl}_2]{}$

41. H—

$\xrightarrow{\text{CH}_3\text{OH, H}^{\oplus}}$

42.

$\xrightarrow[\text{HI}]{\text{excess}}$

43.

$\xrightarrow[\text{acid catalyst}]{\substack{\text{1 equiv.} \\ \text{CH}_3\diagdown \\ \text{CHCH}_2\text{OH} \\ \text{CH}_3\diagup}}$ hemiacetal

44.

$\xrightarrow[\text{ether}]{\text{CH}_3\text{MgBr}}$ $\xrightarrow{\text{H}_3\text{O}^{\oplus}}$

45.

$\xrightarrow[\text{ethanol}]{(\text{CH}_3\text{CH}_2)_2\text{NH}}$

46.

$\xrightarrow{\text{Ph}_3\text{P}=\text{CH}_2}$ $\xrightarrow{\text{H}_3\text{O}^{\oplus}}$

47.

$\xrightarrow{\text{NaBH}_4}$ $\xrightarrow{\text{H}_3\text{O}^{\oplus}}$

48. 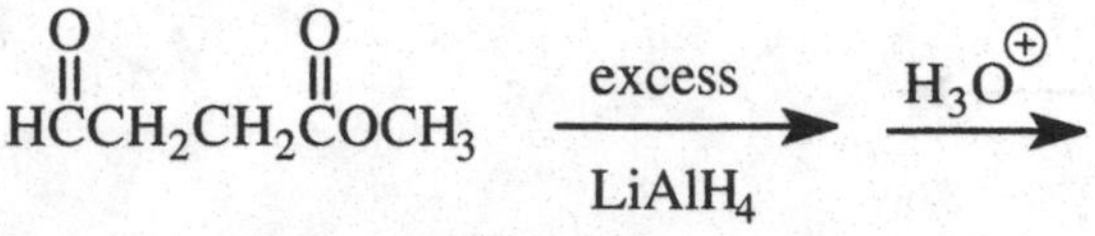 $\overset{O}{\overset{\|}{H}C}CH_2CH_2\overset{O}{\overset{\|}{C}}OCH_3$ $\xrightarrow[\text{LiAlH}_4]{\text{excess}}$ $\xrightarrow{\text{H}_3\text{O}^{\oplus}}$

49. 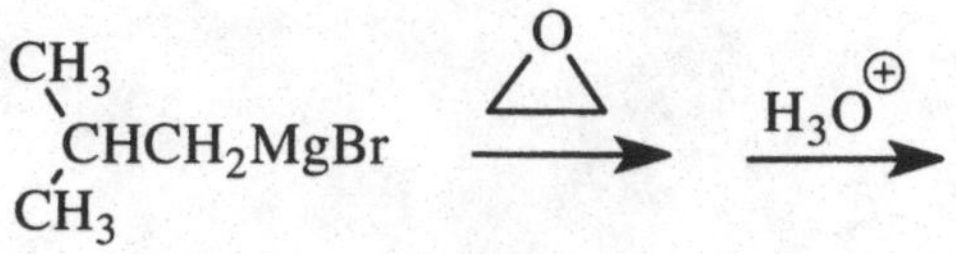 $\overset{CH_3}{\underset{CH_3}{CH}}CHCH_2MgBr$ $\xrightarrow{\text{H}_3\text{O}^{\oplus}}$

50. 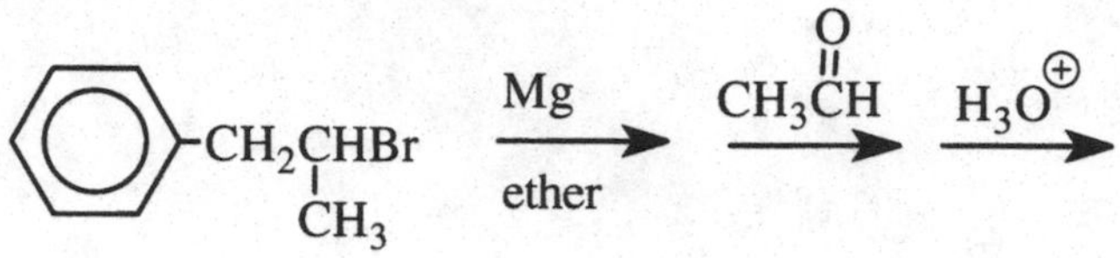 ⬡—$CH_2\underset{CH_3}{CHBr}$ $\xrightarrow[\text{ether}]{\text{Mg}}$ $\xrightarrow{CH_3\overset{O}{\overset{\|}{C}}H}$ $\xrightarrow{\text{H}_3\text{O}^{\oplus}}$

Answers

1. **False**
2. **True**
3. **True**
4. **False**
5. **True**
6. **d**
7. **b**
8. **d**
9. **b**
10. **c**
11. **c**
12. **b**
13. **a**
14. **b**
15. Infrared
16. Nuclear magnetic resonance
17. Infrared
18. Mass spectroscopy
19. Ultraviolet visible
20. Mass spectroscopy
21. Nuclear magnetic resonance
22. Ultraviolet visible

23. 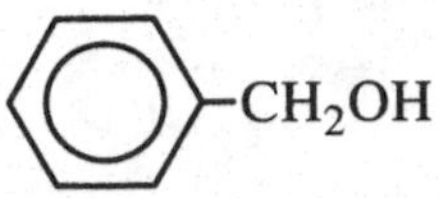 ⟨benzene⟩—CH₂OH

24.  ⟨tetrahydrofuran ring with O⟩

25. $(CH_3CH_2CH_2CH_2)_2S$

26. 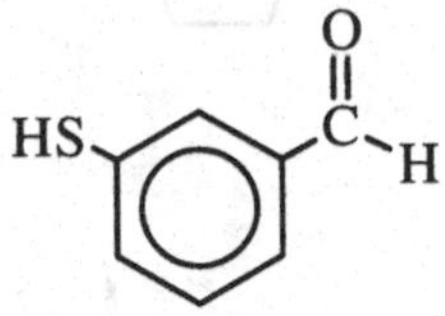 HS—⟨benzene⟩—C(=O)H

27. 7-Decenoic acid

28. 1-Ethylcyclohexanecarboxylic acid

29. 9-Methyl-3-decanol

30. 1,1,3-trimethoxycyclohexane

31. *trans*-2-Heptenal or *E*-2-Heptenal

32. 7,7-Dimethyl-4-octanone

33. 3,4-Diethoxyphenol

34. *m*-Isopropylbenzoic acid or 3-Isopropylbenzoic acid

35. 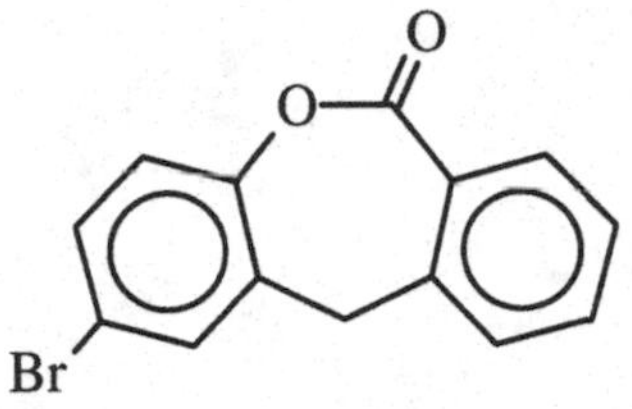 (or ortho to oxygen on same ring)

36. 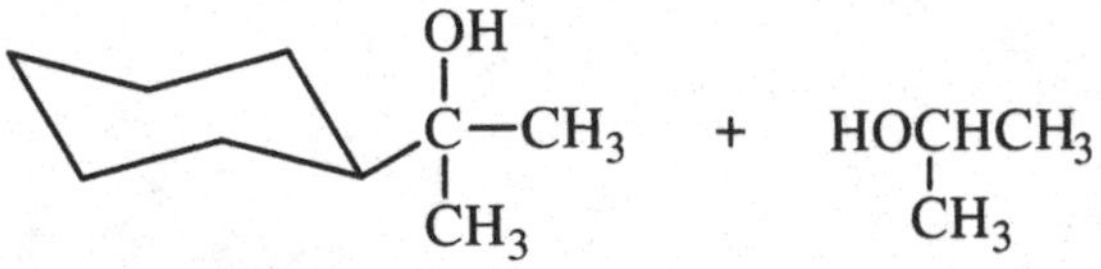 + $HOCHCH_3$ / CH_3

37. 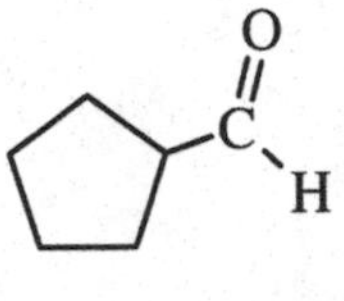   38.

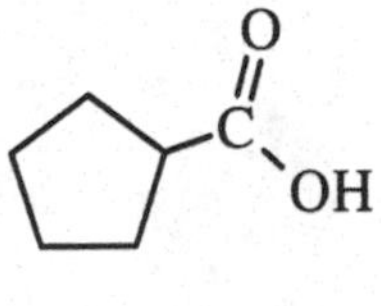

39.

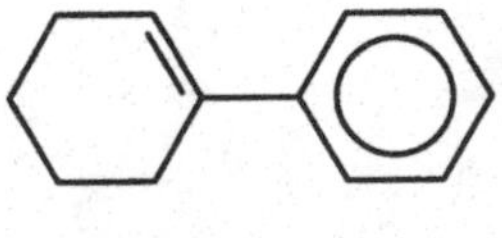

40. 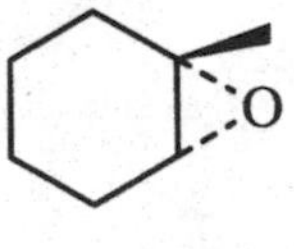 + enantiomer + Cl—⟨benzene⟩—CO_2H

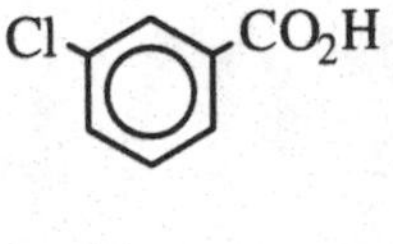

41. 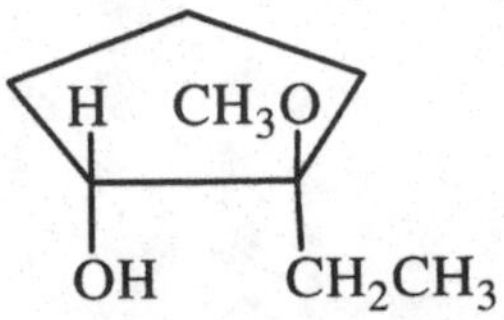   42. $ICH_2CH_2CHCH_2CHCH_3$ + CH_3I (with OH, OH substituents)

43.

44.

45.

46. $+$ $HOCH_2CH_2OH$

47.

48. $HOCH_2CH_2CH_2CH_2OH$ $+$ $HOCH_3$

49. $(CH_3)_2CHCH_2CH_2CH_2OH$

50.

CHAPTER 7

Carboxylic Acid Derivatives: Nucleophilic Acyl Substitution

In this chapter we survey carboxylic acid derivatives with emphasis on nucleophilic acyl substitution. Most derivatives have the general form RCOL, where L is a leaving group. These include everything from the esters (RCO_2R') associated with sweet fragrances of fruits and flowers to amides linkages ($RCONHR'$) that connect amino acids in proteins. Acyl substitutions on compounds of this form are among the most common and important reactions in organic chemistry and biochemistry.

ESSENTIAL BACKGROUND

- **Nucleophiles and leaving groups**
- **Friedel-Crafts acylation (Chapter 1)**
- **Coupling of Gilman reagents with acyl chlorides (Chapter 5)**
- **Carboxylic acid nomenclature (Chapter 6)**
- **Acyl group structure and names (Chapter 6)**
- **Preparation of acyl chlorides and bromides from carboxylic acids (Chapter 6)**

TOPIC 1: STRUCTURES AND NOMENCLATURE

KEY POINTS

✓ *What are acyl halides and how are they named?*

✓ *What are anhydrides and how are they named?*

✓ *What are esters and how are they named?*

✓ *What are amides and how are they classified as primary, secondary, or tertiary?*

✓ *How are amides named?*

✓ *What are nitriles and how are they named?*

As their name implies, **acyl halides** have the general form **RCOX**. Their names are derived from the name of the acid by replacing the "ic acid" with "yl halide." If the acid name ends with "carboxylic acid," that suffix is replaced by "carbonyl halide."

Acetyl chloride   *m*-Bromobenzoyl bromide   1-Cyclobutenecarbonyl fluoride

Loss of water from two carboxylic acid molecules leads to an **anhydride** (means "without water"). **Symmetrical anhydrides** of nonsubstituted monocarboxylic acids or cyclic anhydrides of dicarboxylic acids are named by replacing the word "acid" with "anhydride." For a symmetrical anhydride of a substituted carboxylic acid, the prefix "bis" is used. Unsymmetrical or **mixed anhydrides** are named by alphabetically listing the names of both acids followed by the word "anhydride."

Anhydride

Propanoic anhydride   Bis(trifluoroacetic) anhydride

Succinic Anhydride   Benzoic formic anhydride
(Butanedioic anhydride)

Esters have the general form RCO_2R' and are named by specifying the alkyl group R' followed by the name of the acid RCO_2H with the "ic acid" suffix replaced by "ate."

Phenyl acetate   Methyl benzoate   Isopropyl
1-cyclobutenecarboxylate

Amides have nitrogen attached directly to the carbonyl and are classified as primary, secondary, or tertiary depending on the number of carbons attached to nitrogen. Primary amides are named by replacing the "oic aid" or "ic acid" ending with "amide" or replacing "carboxylic acid" with "carboxamide." Secondary and tertiary amides are named as primary amides bearing substituents on nitrogen (designated with the locator "*N*").

$$CH_3CH-\overset{\overset{\displaystyle O}{\|}}{C}\underset{NH_2}{}$$

2-Methylpropanamide

N-Ethylcyclopentanecarboxamide

N,N-Dimethylformamide
(DMF)
N,N-Dimethylmethanamide

Nitriles contain a C≡N group, and although they are structurally dissimilar from the other acid derivatives (RCOL) above, their chemistry is similar enough that it is convenient to include them here. Simple nitriles are named by adding the suffix "nitrile" to the alkane name where the CN carbon is C1. Replacing the "ic acid" or "oic acid" of the carboxylic acid name with "onitrile" or replacing the suffix "carboxylic acid" with "carbonitrile" names many nitriles.

$$CH_3-\overset{\overset{\displaystyle CH_3}{|}}{\underset{\underset{\displaystyle CH_3}{|}}{C}}-C≡N$$

2,2-Dimethylpropanenitrile   Benzonitrile

CH_3C≡N
Acetonitrile
(Ethanenitrile)

Cyclopentanecarbonitrile

Topic Test 1: Structures and Nomenclature

Provide complete names for the following.

1. [CH_3CH_2CH_2CO]_2O

2. (cyclopentyl)–NHĊCH_2CH_2CH_2CH_2CH_3

3. (CH_3)_3CCH_2CH_2CH_2CH_2C≡N

Provide unambiguous structural formulas for the following compounds named below.

4. Butyl tribromoacetate

5. Bis(4-methylpentanoic) anhydride

6. 1-Cyclobutylcyclohexanecarbonyl fluoride

Topic Test 1: Answers

1. Butanoic anhydride (or butyric anhydride)

2. *N*-cyclopentylhexanamide

3. 6,6-Dimethylheptanenitrile

4. $CBr_3CO_2CH_2CH_2CH_2CH_3$

5. $[(CH_3)_2CHCH_2CH_2CO]_2O$

6.

TOPIC 2: NUCLEOPHILIC ACYL SUBSTITUTION

KEY POINTS

✓ *What are the structural requirements for nucleophilic acyl substitution?*

✓ *What steps are involved in a nucleophilic acyl substitution mechanism?*

As we saw in Chapter 6, the general form of a **nucleophilic acyl substitution** reaction shows some leaving group "L" being replaced by some nucleophile "Nu."

$$RCOL + Nu \rightarrow RCONu + L$$

If the nucleophile is negatively charged, the mechanism is simple **addition–elimination** where the nucleophile attacks the carbonyl to form a tetrahedral intermediate. The carbonyl then reforms as the leaving group simultaneously departs.

If the nucleophile is uncharged, an additional step is required for proton loss.

In some cases the reaction can be **acid catalyzed** and occurs via protonated carbonyls.

$$R-\overset{\overset{O}{\|}}{C}-L \xrightarrow{H^{\oplus}} R-\overset{\overset{\oplus}{O}H}{\underset{\|}{C}}-L \xrightarrow{NuH} R-\overset{\overset{OH}{|}}{\underset{|}{C}}-\overset{\oplus}{N}uH$$

$$\overline{+} H^{\oplus} \downarrow$$

$$R-\overset{\overset{O}{\|}}{C}-Nu \xleftarrow{-H^{\oplus}} R-\overset{\overset{\oplus}{O}H}{\underset{\|}{C}}-Nu \xleftarrow{} R-\overset{:OH}{\underset{\underset{\oplus}{LH}}{C}}-Nu$$

$$+ LH$$

Acyl halides, anhydrides, esters, and amides can each undergo nucleophilic acyl substitution. The reactions differ only in the identity of the nucleophiles and leaving groups. The order of relative reactivity is predictable according to the basicity of the leaving group. A weaker base is a better leaving group and that corresponds to a more reactive RCOL.

$$\text{Leaving group basicity: } X^- < RCOO^- < RO^- < NH_2^-$$

$$\text{Reactivity: } RCOX > RCOOCOR > RCO_2R' > RCONH_2$$

$$\text{Acyl halide > Anhydride > Ester } > \text{Amide}$$

Generally, a less reactive carboxylic acid derivative can be prepared from one that is more reactive using the appropriate nucleophile.

Topic Test 2: Nucleophilic Acyl Substitution

True/False

1. A strong base is a good leaving group in a nucleophilic acyl substitution reaction.

2. Some nucleophilic acyl substitution reactions are acid catalyzed.

Multiple Choice

3. In a nucleophilic acyl substitution reaction
   a. the nucleophile attacks the carbonyl as the leaving group departs simultaneously.
   b. the nucleophile attacks the carbonyl and then the leaving group departs.
   c. the nucleophile attacks the carbonyl after the leaving group departs.
   d. the nucleophile attacks the leaving group, causing it to depart.
   e. None of the above

4. Which compound below will react fastest in a nucleophilic acyl substitution with water?
   a. Acetyl chloride
   b. Acetic anhydride
   c. Ethyl acetate
   d. Acetamide
   e. All of the above

Short Answer

Provide an unambiguous structural formula for the organic product of the following nucleophilic acyl substitution reaction.

5.

$$CH_3\text{-}CH\text{-}CCH_2C\text{-}Cl \atop CH_3 \quad \quad O$$

6. Write a mechanism for the reaction that takes place when benzoyl chloride reacts with ethoxide ion.

Topic Test 2: Answers

1. **False.** If L^- is a strong base, one predicts it will be a reluctant leaving group.

2. **True.** The carbonyl of the reactant is protonated making it more susceptible to nucleophilic attack.

3. **b.** the nucleophile attacks the carbonyl and then the leaving group departs.

4. **a.** Acetyl chloride. In general, acyl chlorides are more reactive with a given nucleophile than are comparable anhydrides, esters, or amides.

5.

6.

TOPIC 3: PREPARATION AND REACTIONS OF ACYL HALIDES

KEY POINTS

✓ *How are acyl chlorides and bromides prepared from carboxylic acids?*

✓ *How are acyl halides converted to other carboxylic acid derivatives?*

✓ *What do the terms hydrolysis, alcoholysis, and aminolysis mean?*

✓ *What reactions from previous chapters involve acyl chlorides?*

✓ *What happens when RCOX reacts with excess Grignard reagent?*

✓ *What reagents will reduce an acyl chloride to a primary alcohol or aldehyde?*

Acyl chlorides and bromides are the most common acyl halides. Recall from Chapter 6 that these are prepared from carboxylic acids with $SOCl_2$ or PBr_3, respectively. Acid chlorides can be converted to anhydrides if a carboxylate is used as the nucleophile.

$$R-\overset{O}{\overset{\|}{C}}-Cl \xrightarrow{Na^{\oplus} \; {}^{\ominus}\overset{O}{\overset{\|}{O}}CR'} R-\overset{O}{\overset{\|}{C}}-O-\overset{O}{\overset{\|}{C}}-R' \quad (+ \; NaCl)$$

Like all the carboxylic acid derivatives in this chapter, acyl halides react with water (**hydrolysis**) to yield carboxylic acids.

$$RCOX + H_2O \rightarrow RCO_2H + HX$$

Similarly, acid halides react with alcohols (**alcoholysis**) or ammonia (**aminolysis**) to yield esters and primary amides, respectively. If monosubstituted or disubstituted amines are used, aminolysis leads to secondary or tertiary amides, respectively. The reaction produces HX so aminolysis usually requires either two equivalents of ammonia or amine (to neutralize the HX) or one equivalent of some other inexpensive base like NaOH.

$$R-\overset{O}{\overset{\|}{C}}-X \quad \xrightarrow[\text{Alcoholysis}]{HOR'} \quad R-\overset{O}{\overset{\|}{C}}-OR'$$

$$\xrightarrow[\text{Aminolysis}]{NH_3} \quad R-\overset{O}{\overset{\|}{C}}-NH_2 \quad (+ \; NH_4^{\oplus} X^{\ominus})$$

$$\xrightarrow[\text{NaOH (aq)}]{NH_2R' \text{ or } NHR'_2} \quad R-\overset{O}{\overset{\|}{C}}-NHR' \quad \text{or} \quad R-\overset{O}{\overset{\|}{C}}-NR'_2$$

$$(+ \; NaX + H_2O)$$

Recall that acyl chlorides can be converted to ketones by coupling with Gilman reagents (Chapter 5) or to phenones via the Friedel-Crafts acylation (Chapter 1).

$$H-Ar \quad \xrightarrow[\text{AlCl}_3]{R-\overset{O}{\overset{\|}{C}}-Cl} \quad R-\overset{O}{\overset{\|}{C}}-Ar$$

$$R-\overset{O}{\overset{\|}{C}}-Cl \quad \xrightarrow{R'_2CuLi} \quad R-\overset{O}{\overset{\|}{C}}-R'$$

Acyl halides react with two equivalents of Grignard reagent to yield tertiary alcohols.

$$R-\overset{O}{\overset{\|}{C}}-Cl \quad \xrightarrow[\text{2) } H_3O^{\oplus}]{\text{1) } 2 \, R'MgX} \quad R-\overset{R'}{\underset{R'}{\overset{|}{\underset{|}{C}}}}-OH$$

Reduction of acyl chlorides can be carried out with lithium aluminum hydride to yield primary alcohols. If one equivalent of the milder reducing agent lithium tri-*tert*-butoxyaluminum hydride is used, the reduction stops at the aldehyde stage.

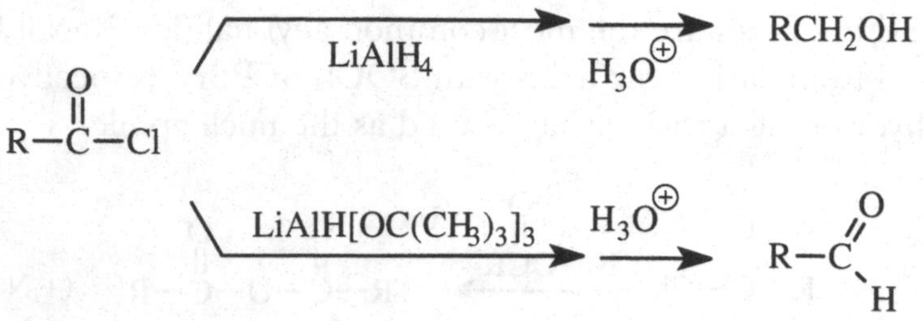

LiAlH4
H3O⊕
RCH2OH
R—C—Cl
O
LiAlH[OC(CH3)3]3
H3O⊕
R—C
O
H

Topic Test 3: Preparation and Reactions of Acyl Halides

True/False

1. Hydrolysis of an acyl chloride yields an ester.
2. An acyl halide will react with one equivalent of Gilman reagent to yield a ketone.

Multiple Choice

3. Which product results from the reaction of propanoyl chloride with two equivalents of CH_3CH_2MgBr followed by aqueous acid?
   a. 2-Pentanone
   b. 3-Ethyl-3-pentanol
   c. Propanoic acid + acetic acid
   d. Acetic anhydride
   e. None of the above
4. Which reagents below will convert benzoyl chloride into benzaldehyde?
   a. PhMgBr, then H_3O^+
   b. $AlCl_3$
   c. Lithium tri-*tert*-butoxyaluminum hydride, then H_3O^+
   d. Lithium aluminum hydride, then H_3O^+
   e. None of the above

Short Answer

Complete the following reaction schemes with unambiguous structural formulas for the missing organic compounds.

5. 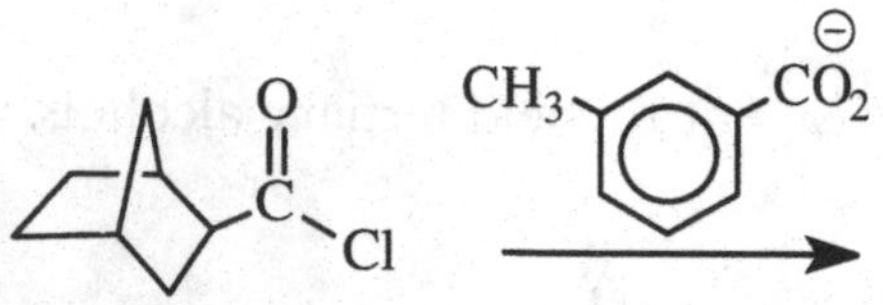

O
C
Cl
CH3
CO2⊖

6. 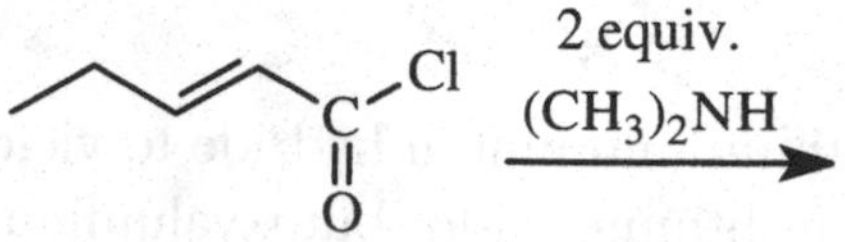

Cl
C
O
2 equiv.
(CH3)2NH

Topic Test 3: Answers

1. **False.** Hydrolysis of an acyl halide produces a carboxylic acid.

2. **True.** R'COCl + LiCuR$_2$ → R'COR (+ CuR + LiCl).

3. **b.** 3-Ethyl-3-pentanol. Note that the product is a tertiary alcohol in which two groups around the alcohol carbon came from the Grignard reagent: (CH$_3$CH$_2$)$_3$COH.

4. **c.** Lithium tri-*tert*-butoxyaluminum hydride then H$_3$O$^+$.

5.

6.

TOPIC 4: PREPARATION AND REACTIONS OF ANHYDRIDES

KEY POINTS

✓ *How are anhydrides prepared?*

✓ *What products result from hydrolysis of an anhydride?*

✓ *What products result from alcoholysis of an anhydride?*

✓ *What products result from aminolysis of an anhydride?*

✓ *What products result from LiAlH$_4$ reduction of an anhydride?*

Although anhydrides result from dehydration of carboxylic acids with heat or P$_2$O$_5$, this strategy is limited to symmetrical or cyclic anhydrides. The more general method involves treatment of an acyl chloride with a carboxylate (Topic 3).

Anhydrides undergo hydrolysis, alcoholysis, and aminolysis to yield carboxylic acids, esters, and amides, respectively. Reduction of anhydrides with lithium aluminum hydride yields primary alcohols.

The reaction scheme at top:

$R-\overset{\overset{O}{\|}}{C}-O-\overset{\overset{O}{\|}}{C}-R$

$\xrightarrow{H_2O}$ 2 RCO_2H

$\xrightarrow{HOR'}$ $R-\overset{\overset{O}{\|}}{C}-OR'$ + $HO-\overset{\overset{O}{\|}}{C}-R$

$\xrightarrow{2\ NH_3}$ $R-\overset{\overset{O}{\|}}{C}-NH_2$ + $[\overset{\oplus}{NH_4}][\overset{\ominus}{O_2}CR]$

$\xrightarrow[\text{ether}]{LiAlH_4}$ $\xrightarrow{H_3O^{\oplus}}$ 2 RCH_2OH

Topic Test 4: Preparation and Reactions of Anhydrides

True/False

1. Hydrolysis of an anhydride yields two equivalents of carboxylic acid.

2. Reaction of acetic anhydride with excess CH_3NH_2 will yield N-methylacetamide.

Multiple Choice

3. Which product results from the reaction sequence shown?

$\xrightarrow{LiAlH_4}$ $\xrightarrow{H_3O^{\oplus}}$

   a. Two equivalents of $CH_3CH_2CH_2CH_2CH_2OH$
   b. 1,3-Cyclohexanediol
   c. $HOOC(CH_2)_3COOH$
   d. $HOCH_2CH_2CH_2CH_2CH_2OH$
   e. None of the above

4. Which pair of reactants below will yield butanoic anhydride?
   a. Butanoyl chloride and 1-butanol
   b. Butanoic acid and water
   c. Butanoyl chloride and sodium butanoate
   d. All of the above
   e. None of the above

Short Answer

5. $\xrightarrow{CH_3CH_2OH}$

6.

Topic Test 4: Answers

1. **True.** This is the reverse of the dehydration reaction sometimes used to prepare symmetrical anhydrides.

2. **True.** This is aminolysis using a methyl-substituted ammonia (called methylamine) as the nucleophile to produce a secondary amide, $CH_3CONHCH_3$. Also produced is the salt, methylammonium acetate, $[CH_3NH_3^+][CH_3CO_2^-]$.

3. **d.** $HOCH_2CH_2CH_2CH_2CH_2OH$

4. **c.** Butanoyl chloride and sodium butanoate. Reaction of butanoyl chloride with 1-butanol (a) would yield the ester butyl butanoate. Butanoic acid and water (b) do not react except for the modest acid dissociation to give some butanoate and hydronium (Chapter 6).

5.

6.

TOPIC 5: PREPARATION AND REACTIONS OF ESTERS

KEY POINTS

✓ *How are esters prepared from acyl halides or anhydrides?*

✓ *What are the reactants, products, and mechanism for the Fischer esterification?*

✓ *What are the products of ester hydrolysis or aminolysis?*

✓ *What are the reactants and products of ester saponification?*

✓ *How can esters be reduced to alcohols or aldehydes?*

We saw above that esters can be prepared by alcoholysis of acyl halides or anhydrides.

Many esters can be prepared by acid-catalyzed reaction of a carboxylic acid with an alcohol. The process is often called the **Fischer esterification**. The reversible equilibrium can be favored in either direction (Le Châtelier's principle). The reverse process is acid-catalyzed ester hydrolysis.

$$R-\overset{\overset{\textstyle O}{\|}}{C}-OH \ + \ HOR' \underset{\underset{\text{Acid-catalyzed}}{\underset{\text{hydrolysis}}{H^+}}}{\overset{\overset{\text{Fischer}}{\overset{\text{esterification}}{\longrightarrow}}}{\rightleftarrows}} R-\overset{\overset{\textstyle O}{\|}}{C}-OR' \ + \ H_2O$$

The alkaline hydrolysis of esters is called **saponification** and leads to carboxylates and alcohols.

$$R-\overset{\overset{\textstyle O}{\|}}{C}-OR' \ \xrightarrow[\text{H}_2\text{O}]{\text{OH}^{\ominus}} \ R-CO_2^{\ominus} \ + \ HOR'$$

Aminolysis of esters leads to amides.

$$R-\overset{\overset{\textstyle O}{\|}}{C}-OR' \ \xrightarrow{\text{NH}_3} \ R-\overset{\overset{\textstyle O}{\|}}{C}-NH_2 \ + \ HOR'$$

$$\xrightarrow{\text{NH}_2\text{R}} \ R-\overset{\overset{\textstyle O}{\|}}{C}-NHR \ + \ HOR'$$

$$\xrightarrow{\text{NHR}_2} \ R-\overset{\overset{\textstyle O}{\|}}{C}-NR_2 \ + \ HOR'$$

Recall from Chapter 3 that esters can be reduced to alcohols with lithium aluminum hydride. If the reduction is carried out with one equivalent of diisobutylaluminum hydride (DIBAH, also called DIBAL-H), an aldehyde is formed.

$$R-\overset{\overset{\textstyle O}{\|}}{C}-OR' \ \xrightarrow[\text{ether}]{\text{LiAlH}_4} \ \xrightarrow{\text{H}_3\text{O}^{\oplus}} \ RCH_2OH \ + \ HOR'$$

$$\xrightarrow[-78°\text{ C, toluene}]{\underset{[(CH_3)_2CHCH_2]_2AlH}{1 \text{ equivalent}}} \ \xrightarrow{\text{H}_3\text{O}^{\oplus}} \ R-\overset{\overset{\textstyle O}{\|}}{C}-H \ + \ HOR'$$

Reaction of an ester with excess Grignard reagent yields alcohols (Chapter 3).

$$R-\overset{\overset{\textstyle O}{\|}}{C}-OR' \ \xrightarrow[\text{ether}]{2 \ R''\text{MgX}} \ \xrightarrow{\text{H}_3\text{O}^{\oplus}} \ R-\overset{\overset{\textstyle R''}{|}}{\underset{\underset{\textstyle R''}{|}}{C}}-OH \ + \ HOR'$$

Topic Test 5: Preparation and Reactions of Esters

True/False

1. The acid-catalyzed hydrolysis of an ester is the reverse of the Fischer esterification.

2. Esters can be prepared by alcoholysis of anhydrides.

Multiple Choice

3. Which products result from the reaction $CH_3CH_2COOCH_3$ with H_2O, NaOH?
   a. $CH_3CH_2COOH + HOCH_3$
   b. $CH_3CH_2COO^-Na^+ + HOCH_3$
   c. $CH_3CH_2CH_2OH + HOCH_3$
   d. Propanoic anhydride + $NaOCH_3$
   e. None of the above

4. Which reaction below will lead to an ester?
   a. Reaction of pentanoic acid with ethanol in the presence of dry acid catalyst
   b. Reaction of pentanoyl chloride with ethanol
   c. Reaction of pentanoic anhydride with ethanol
   d. All of the above
   e. None of the above

Short Answer

Provide unambiguous structural formulas for the missing organic compounds.

5. $\xrightarrow{\text{LiAlH}_4} \xrightarrow{\text{H}_3\text{O}^{\oplus}}$

6. $\xrightarrow[\text{-78° C}]{\text{DIBAH}} \xrightarrow{\text{H}_3\text{O}^{\oplus}}$

Topic Test 5: Answers

1. **True.**

2. **True.** $RCOOCOR + HOR' \rightarrow RCOOR' + HOOCR$

3. **b.** $CH_3CH_2COO^-Na^+ + HOCH_3$. This is saponification, also called alkaline ester hydrolysis. The carboxylate is often shown without the sodium cation: $CH_3CH_2COO^-$.

4. **d.** All of the above. Choices a, b, and c are the Fischer esterification and alcoholysis of an acyl chloride or anhydride, respectively.

5. $HOCH_2CH_2CH_2CH_2CH_2OH$

6. $O{=}CHCH_2CH_2CH_2CH_2OH$

TOPIC 6: PREPARATION AND REACTIONS OF AMIDES

KEY POINTS

✓ *How are amides prepared?*

✓ *What are the products of amide hydrolysis?*

✓ *What products result from lithium aluminum hydride reduction of amides?*

✓ *How can primary amides be dehydrated to nitriles?*

Amides are most often prepared by aminolysis of acyl halides, anhydrides, or esters as we saw in the previous sections. The low reactivity of amides means they undergo fewer nucleophilic acyl substitution reactions than the other carboxylic acid derivatives we have discussed. Hydrolysis occurs when amides are heated under acidic or alkaline conditions.

$$R-\overset{\overset{\displaystyle O}{\|}}{C}-\underset{|}{N}- \xrightarrow{H_3O^{\oplus}} RCO_2H \;+\; H_2\overset{\oplus}{\underset{|}{N}}-$$

$$\xrightarrow{H_2O, \; OH^{\ominus}} \left[RCO_2^{\ominus} \;+\; \underset{|}{N}H- \right] \xrightarrow{H_3O^{\oplus}} RCO_2H \;+\; H_2\overset{\oplus}{\underset{|}{N}}-$$

Lithium aluminum hydride reduction of amides will convert the carbonyl to a methylene, CH_2. The resulting product will be an amine bearing at least one primary alkyl group.

$$R-\overset{\overset{\displaystyle O}{\|}}{C}-\underset{|}{N}- \xrightarrow{LiAlH_4} \xrightarrow{H_2O} R-CH_2-\underset{|}{N}-$$

Primary amides can undergo an unusual reaction that is not a nucleophilic acyl substitution. Any of several dehydrating agents can be used to convert the primary amide into a nitrile.

$$R-\overset{\overset{\displaystyle O}{\|}}{C}-NH_2 \xrightarrow[\text{or } P_2O_5 \text{ or } POCl_3]{SOCl_2, \text{ benzene, heat}} R-C\equiv N$$

$$\text{(loss of } H_2O)$$

Topic Test 6: Preparation and Reactions of Amides

True/False

1. Acetamide reacts quickly with water at room temperature to give acetic acid and ammonia.

2. Reduction of an amide with lithium aluminum hydride then water will yield a primary alcohol.

Multiple Choice

3. Which is the product of the reaction below?

$\xrightarrow{\text{LiAlH}_4}$ $\xrightarrow{\text{H}_3\overset{\oplus}{\text{O}}}$

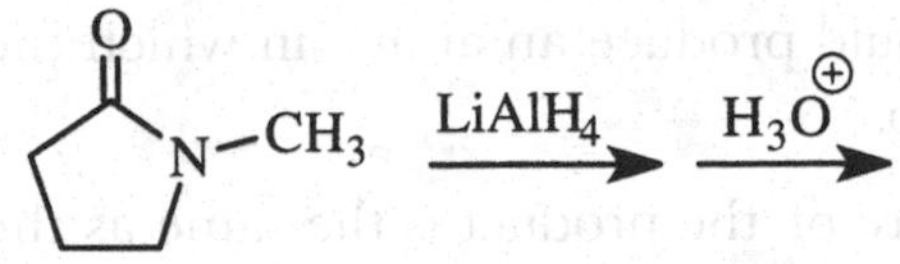

a.

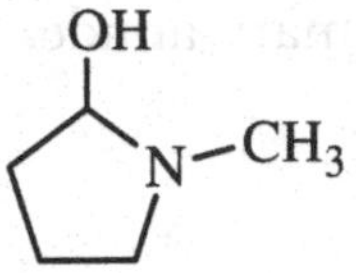

b.

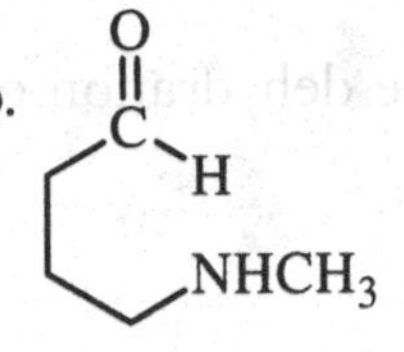

c.

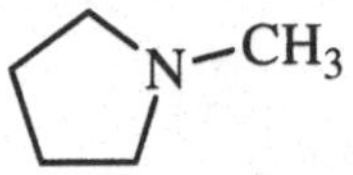

d. $\text{HOCH}_2\text{CH}_2\text{CH}_2\text{CH}_2\text{NHCH}_3$

e. None of these

4. Which is the product of the reaction below?

$\xrightarrow[\substack{\text{benzene} \\ 80°\text{ C}}]{\text{SOCl}_2}$

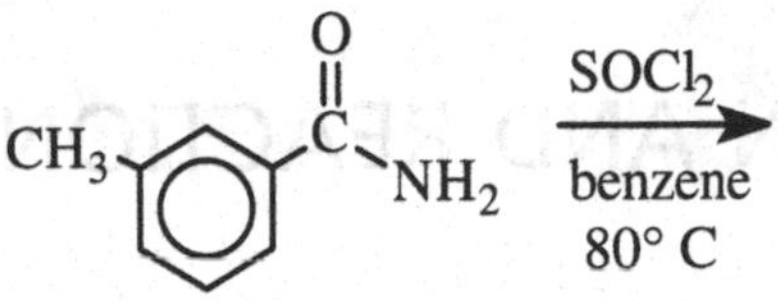

a.

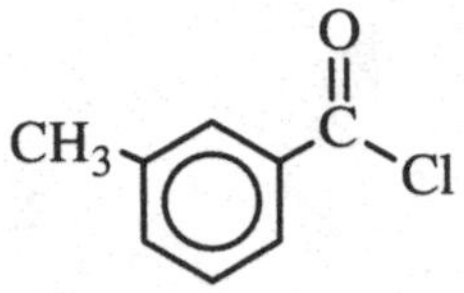

b.

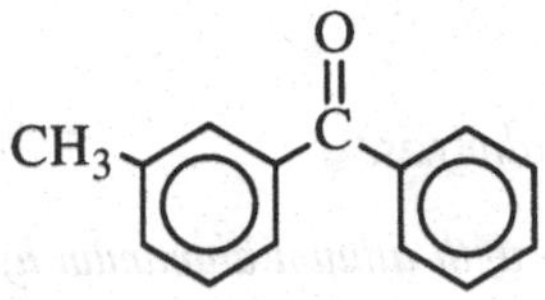

c.

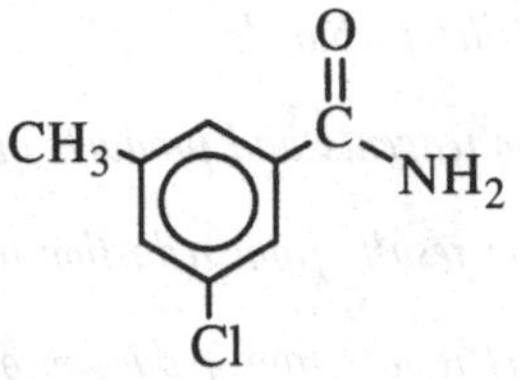

d.

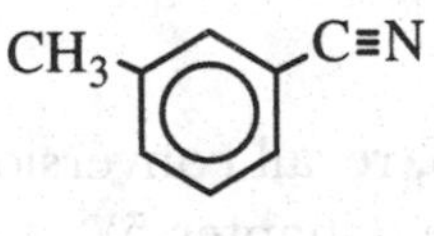

e. None of these

Short Answer

Provide unambiguous structural formulas for the missing organic compounds.

5. 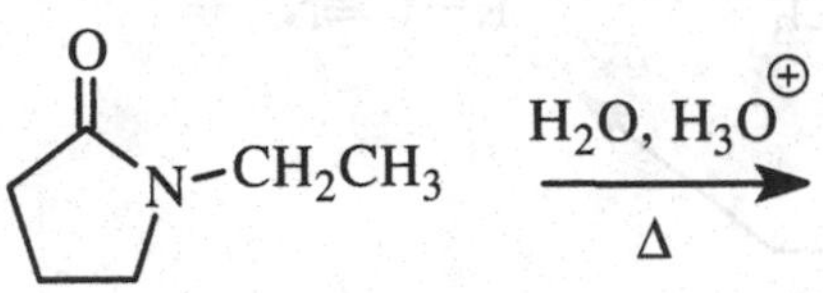 $\xrightarrow[\Delta]{\text{H}_2\text{O, H}_3\overset{\oplus}{\text{O}}}$

6. $\text{CH}_3\text{CH}_2\overset{\displaystyle\overset{\text{O}}{\|}}{\text{C}}\text{NCH}_2\text{CH}_2\text{CH}_2\text{CH}_3$ $\xrightarrow[\Delta]{\text{NaOH, H}_2\text{O}}$

   with CH_3 on N

Topic Test 6: Answers

1. **False.** Although the hydrolysis of acetamide would yield acetic acid, this reaction is difficult and requires elevated temperatures and acid or base.

2. **False.** The reaction would produce an amine in which the carbonyl of the amide is reduced to a CH_2 group.

3. **c.** Note that the structure of the product is the same as the starting cyclic amide (called a lactam) except that the carbonyl has been reduced to a methylene group, CH_2.

4. **d.** *m*-Methylbenzonitrile results from the dehydration of the primary amide.

5. $HO\overset{\overset{O}{\|}}{C}(CH_2)_3\overset{\oplus}{N}H_2CH_2CH_3$

6. $CH_3CH_2CO_2^{\ominus}$ + $CH_3CH_2CH_2CH_2NHCH_3$

TOPIC 7: PREPARATION AND REACTIONS OF NITRILES

KEY POINTS

✓ *How are nitriles prepared?*

✓ *What are the reagents and products of nitrile hydrolysis?*

✓ *What product results from reduction of a nitrile with lithium aluminum hydride?*

✓ *What product results from reduction of a nitrile with DIBAH?*

✓ *What product results from reaction of a nitrile with a Grignard reagent?*

In addition to dehydration of primary amides discussed in Topic 6, recall conversion of alkyl halides to nitriles by S_N2 reactions using cyanide as the nucleophile (Chapter 5).

Nitriles undergo partial hydrolysis in aqueous acid or base to yield amides. Amides are, of course, also susceptible to further hydrolysis. Complete hydrolysis results if the nitrile is heated more vigorously in aqueous acid or base.

$$R-C\equiv N \xrightarrow[H_3O^{\oplus}]{H_2O} R-\overset{\overset{\displaystyle O}{\|}}{C}-NH_2 \xrightarrow[H_3O^{\oplus}]{H_2O} RCO_2H + NH_4^{\oplus}$$

$$\downarrow \begin{array}{c} H_2O \\ OH^{\ominus} \end{array}$$

$$R-\overset{\overset{\displaystyle O}{\|}}{C}-NH_2 \xrightarrow[OH^{\ominus}]{H_2O} \left[RCO_2^{\ominus} + NH_3 \right] \xrightarrow{H_3O^{\oplus}}$$

Reduction of a nitrile with lithium aluminum hydride then water yields a saturated product. If one equivalent of DIBAH is used as the hydride source, an imine anion is produced that undergoes hydrolysis to yield an aldehyde. If an alkyl anion (from a Grignard reagent) is used in place of the hydride, a ketone results.

$$R-C\equiv N \begin{cases} \xrightarrow{LiAlH_4} \xrightarrow{H_2O} RCH_2NH_2 \\[2ex] \xrightarrow[\substack{Toluene \\ -78^\circ C}]{DIBAH} \left[R-\overset{\overset{\displaystyle N^{\ominus}}{\|}}{C}-H \right] \xrightarrow{H_2O} R-\overset{\overset{\displaystyle O}{\|}}{C}-H \\[2ex] \xrightarrow[ether]{R'MgX} \left[R-\overset{\overset{\displaystyle N^{\ominus}}{\|}}{C}-R' \right] \xrightarrow{H_2O} R-\overset{\overset{\displaystyle O}{\|}}{C}-R' \end{cases}$$

Topic Test 7: Preparation and Reactions of Nitriles

True/False

1. Nitriles can be hydrolyzed by heating in aqueous acid to yield carboxylic acids.
2. 2,2-Dimethylpropanenitrile can be prepared by S_N2 reaction of cyanide ion on 2-bromo-2-methylpropane.

Multiple Choice

3. Which reagents below will convert $CH_3CH_2C\equiv N$ to propanal?
   a. $LiAlH_4$, then water
   b. $[(CH_3)_2CHCH_2]_2AlH$ in toluene, -78°, then water
   c. CH_3CH_2MgBr, then H_3O^+
   d. NaOH, water and heat
   e. None of the above

4. Which product below results from treatment of acetonitrile with lithium aluminum hydride then water?
   a. Acetaldehyde, $CH_3CH=O$
   b. Ethanol, CH_3CH_2OH
   c. Ethylamine, $CH_3CH_2NH_2$
   d. $CH_3CH=NH$
   e. None of the above

Short Answer

5. Name the product that results from treating benzamide with P_2O_5.

6. Provide an unambiguous structural formula for the missing organic compound.

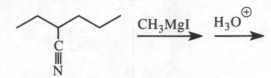

$$\xrightarrow{CH_3MgI} \xrightarrow{H_3O^{\oplus}}$$

Topic Test 7: Answers

1. **True.** $RC\equiv N \rightarrow\rightarrow RCOOH$

2. **False.** The proposed starting material, $(CH_3)_3C$—Br is a tertiary alkyl halide and is not suitable for S_N2 reactions.

3. **b.** $[(CH_3)_2CHCH_2]2AlH$, then water. The reducing agent DIBAH adds one equivalent of hydride to the nitrile carbon to yield an imine anion that is protonated and hydrolyzed to the aldehyde upon reaction with water.

4. **c.** Ethylamine, $CH_3CH_2NH_2$

5. Benzonitrile. This is a dehydration reaction: $PhCONH_2 + P_2O_5 \rightarrow Ph$—$C\equiv N$.

6.

APPLICATION

Polyamides: Proteins and Nylons

Alpha amino acids (usually simply called **amino acids**) may be generalized as shown below. There are about 20 of these that are commonly encountered, and they differ by the identity of R. Secondary amide linkages can be formed between the amine of one amino acid and the carbonyl of another such that amino acids can be strung together in chains. These secondary amide linkages are known to biochemists as **peptide bonds**. Polymers of amino acids are called **polypeptides** or **proteins**. Proteins are remarkably diverse and are involved in a great number of cellular events.

Peptide bonds

Synthetic polyamides are commonly called nylons. The material nylon 66, for example, can be prepared by the reaction of the six-carbon diacyl chloride with a six-carbon diamine.

$$Cl-\overset{\overset{O}{\|}}{C}-(CH_2)_4-\overset{\overset{O}{\|}}{C}-Cl \quad + \quad H_2N(CH_2)_6NH_2$$

Hexanedionyl chloride   1,6-Diaminohexane
(adipoyl chloride)   (hexamethylenediamine)

$$\left(\overset{\overset{O}{\|}}{C}-(CH_2)_4-\overset{\overset{O}{\|}}{C}-NH-(CH_2)_6-NH\right)_n \qquad \text{Nylon 66}$$

DEMONSTRATION PROBLEM

Show how one could synthesize the amine below using butanoic acid and aniline as the only sources of carbon. More than one step is required.

$$CH_3CH_2CH_2CH_2NH-\bigcirc$$

Solution

Recall that multistep synthesis problems are usually easier to analyze backward. The disubstituted amine product must have come from the analogous amide that appears to have resulted from nucleophilic acyl substitution on butanoyl chloride with aniline. A reasonable alternative is aminolysis of butanoic anhydride.

$$CH_3CH_2CH_2CH_2NH-\bigcirc \xleftarrow[\text{H}_3\text{O}^{\oplus}]{\text{LiAlH}_4} CH_3CH_2CH_2\overset{\overset{O}{\|}}{C}-NH-\bigcirc$$

$$NH_2-\bigcirc$$

$$CH_3CH_2CH_2CO_2H \xrightarrow{SOCl_2} CH_3CH_2CH_2\overset{\overset{O}{\|}}{C}Cl$$

$$\xrightarrow{P_2O_5}$$

Chapter Test
True/False

1. Amides are generally less reactive than esters for nucleophilic acyl substitution with a given nucleophile.

2. Saponification of pentyl acetate will yield acetate and 1-pentanol.

3. Carboxylate, RCO_2^-, is a better leaving group than NH_2^-.

4. *N*-Propylbutanamide is an example of a primary amide.

5. Treatment of bromobenzene with NaCN will yield benzonitrile.

Multiple Choice

6. Which compound below will not undergo hydrolysis to yield acetic acid?
   a. $CH_3C\equiv N$
   b. Ethanoic anhydride
   c. Acetaldehyde, $CH_3CH\!=\!O$
   d. Acetyl chloride
   e. All of the above

7. To make an ester by Fischer esterification, one would begin with
   a. an alcohol and an acyl chloride.
   b. an alcohol and an anhydride.
   c. an alcohol and an amide.
   d. an alcohol and a carboxylic acid.
   e. All of the above

8. What nucleophile below would be most likely used for the synthesis of *N*,*N*-dimethylbenzamide from benzoic anhydride?
   a. NH_3
   b. NH_2CH_3
   c. $NH(CH_3)_2$
   d. $N(CH_3)_3$
   e. None of the above

9. Which reagents will convert propanoic acid to propanoic anhydride?
   a. H_2O, H_3O^+
   b. $SOCl_2$, then $NaO_2CCH_2CH_3$
   c. $CH_3CH_2CH_2MgBr$, then H_3O^+
   d. All of the above
   e. None of the above

10. What is produced when propanamide is heated with aqueous acid?
    a. CH_3CH_2COOH and NH_4^+
    b. $CH_3CH_2CH_2NH_2$
    c. $CH_3CH_2C\equiv N$
    d. $CH_3CH_2CONH_3^+$
    e. None of the above

Short Answer

Complete the following reaction schemes with unambiguous structural formulas for the missing organic compounds.

11. 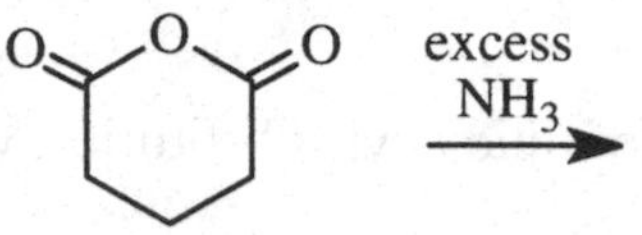 $\xrightarrow[\text{NH}_3]{\text{excess}}$

12.
$$
\begin{array}{c}
\text{CH}_3 \\
\text{CHCH}_2\text{Br} \\
\text{CH}_3
\end{array}
\xrightarrow[\text{DMF}]{\text{NaCN}}
\xrightarrow[\text{ether}]{\text{CH}_3\text{MgI}}
\xrightarrow{\text{H}_3\text{O}^{\oplus}}
$$

13.

$$\xrightarrow{\text{H}_2\text{O}} \xrightarrow{\text{SOCl}_2}$$

14.

$$\xrightarrow{\text{CH}_3\overset{\text{O}}{\overset{\|}{\text{C}}}\text{O}\overset{\text{O}}{\overset{\|}{\text{C}}}\text{CH}_3}$$

15.

$$\xrightarrow[\Delta]{\text{KOH, H}_2\text{O}}$$

Provide structural formulas for the product(s) of each reaction scheme described.

16. Cyclohexanecarbonyl chloride is treated with excess CH_3MgBr, then aqueous acid.

17. Isopropyl butanoate is treated with DIBAH at $-78°\text{C}$ in toluene, then aqueous acid.

18. 2-Methylpentanamide is treated with P_2O_5.

19. CF_3COOH is treated with P_2O_5.

20. Propanenitrile is treated with lithium aluminum hydride then water.

Chapter Test: Answers

1. **True**
2. **True**
3. **True**
4. **False**
5. **False**
6. **c**   7. **d**   8. **c**   9. **b**   10. **a**

11. $\text{H}_2\text{N}\overset{\text{O}}{\overset{\|}{\text{C}}}(\text{CH}_2)_3\text{CO}_2^{\ominus}\ \text{NH}_4^{\oplus}$

12. $\begin{array}{c}\text{CH}_3 \\ \text{CHCH}_2\overset{\text{O}}{\overset{\|}{\text{C}}}\text{CH}_3 \\ \text{CH}_3\end{array}$

13.

14.
$$CH_3\overset{O}{\overset{\|}{C}} \overset{O}{\underset{}{}} CH_2O\text{-}\overset{O}{\overset{\|}{C}}CH_3 \quad + \quad CH_3CO_2H$$

15.
$$\overset{CO_2^{\ominus}K^{\oplus}}{\bigcirc} \quad + \quad HN(CH_2CH_2CH_3)_2$$

16.
$$\overset{OH}{\bigcirc}$$

17. $CH_3CH_2CH_2\overset{O}{\overset{\|}{C}}\text{-}H \quad + \quad HOCHCH_3$
$$\underset{CH_3}{}$$

18.
$$\overset{CH_3}{\underset{|}{}}$$
$$CH_3CH_2CH_2CHC{\equiv}N$$

19.
$$CF_3\overset{O}{\overset{\|}{C}}O\overset{O}{\overset{\|}{C}}CF_3$$

20. $CH_3CH_2CH_2NH_2$

Check Your Performance

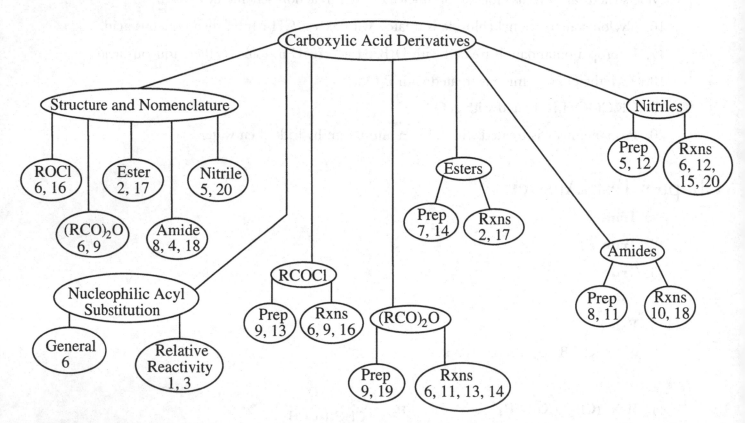

Note the number of questions in each grouping that you got wrong on the chapter test. Identify areas where you need further review and go back to relevant parts of this chapter.

CHAPTER 8

Carbonyl Alpha Substitution and Condensation Reactions

In this chapter we survey reactions that take place on the position adjacent to the carbonyl. Among the most important of these are reactions in which two carbonyl-containing molecules undergo a **condensation reaction** (i.e., they are combined accompanied by a loss of some small molecule such as water or an alcohol). Transformations of this form are found throughout biochemistry.

ESSENTIAL BACKGROUND

- Resonance
- Nucleophiles and electrophiles
- Keto-enol tautomerism
- S_N2 reactions of alkyl halides
- Addition to carbonyl groups (Chapter 5)
- Enamine formation from amines and ketones or aldehydes (Chapter 5)
- Conjugate addition (Chapter 5)
- Preparation of acyl bromides from carboxylic acids with PBr_3 (Chapters 6 and 7)
- Hydrolysis of acyl halides to carboxylic acids (Chapter 7)
- Nomenclature of aldehydes, ketones, carboxylic acids, and their derivatives (Chapters 5–7)

TOPIC 1: REACTIVITY OF ALPHA HYDROGENS— ENOLS AND ENOLATES

KEY POINTS

✓ *What is an alpha hydrogen?*

✓ *What is an enol?*

✓ *How do enols react with electrophilic species?*

✓ *What is an enolate?*

✓ *How do enols react with electrophilic species?*

✓ *What is the general form of an alpha substitution reaction?*

An **alpha hydrogen** is a hydrogen atom on a carbon adjacent to carbonyl. An alpha hydrogen is unusually reactive compared with many other types of C—H. It can be removed under acidic conditions to form an **enol** (recall keto-enol tautomerism).

Alpha Hydrogen

Keto tautomer   Enol tautomer

Enol formation is reversible, and most molecules are in the keto form at any one time. Enol nucleophilicity is predictable. The mechanism for going from the enol to the keto form involves the alpha carbon acting as a nucleophile to capture a proton. There is no requirement that the original alpha proton go back on. It could be a different proton or even an entirely different electrophile, E^+, that might be present.

An alpha hydrogen can be removed by base to form a resonance-stabilized **enolate** (i.e., conjugate base of an enol). Note the partial negative charge on the alpha carbon of an enolate. That position is nucleophilic and can attack an electrophile.

When the alpha hydrogen is replaced by some electrophile, the net transformation is called an **alpha substitution** reaction. Although the carbonyl appears unchanged by the reaction, its presence is essential for the mechanism.

Topic Test 1: Reactivity of Alpha Hydrogens— Enols and Enolates

True/False

1. Benzaldehyde contains one alpha hydrogen.

2. An enol is negatively charged.

Multiple Choice

3. When acetophenone ($PhCOCH_3$) undergoes an alpha substitution reaction with some electrophile, E^+, the product will be

a.

b.

c.

d.

e. None of these

4. Which compound below will undergo an alpha substitution reaction?
   a. Cyclohexanone
   b. Methanal (formaldehyde)
   c. 2,2,4,4-Tetramethyl-3-pentanone
   d. All of the above
   e. None of the above

Short Answer

5. Draw the enol tautomer that is in equilibrium with cyclopentanone.

6. Draw all reasonable resonance forms of the enolate that results from treating cyclopentanone with one equivalent of strong base.

Topic Test 1: Answers

1. **False.** Benzaldehyde, $PhCH{=}O$ has *no* alpha hydrogens.

2. **False.** An enol is electrically neutral and is the tautomer of a carbonyl compound that has at least one alpha hydrogen.

3. **c.** $PhCOCH_2E$ results from replacing one of the only three alpha hydrogens of the starting material with E.

4. **a.** Cyclohexanone. This is the only choice that contains the alpha hydrogens required for an alpha substitution reaction.

5. There are four equivalent alpha hydrogens in cyclopentanone and any one of them can be removed to give the same enol.

6.

There should be no more than two resonance forms shown in your answer and a double-headed arrow should separate them. Recall that resonance forms are different pictures of the same thing and therefore must have the same atoms in the same places.

TOPIC 2: ALPHA SUBSTITUTION REACTIONS

KEY POINTS

✓ *How can aldehydes and ketones be alpha halogenated?*

✓ *How can carboxylic acids be alpha brominated?*

✓ *What is the haloform reaction?*

✓ *How can alkyl groups be placed on an alpha carbon?*

Aldehydes and ketones undergo alpha halogenation under acidic conditions. The reaction is via an enol nucleophile and the electrophile is molecular halogen.

In the **Hell-Volhard-Zelinskii (HVZ) reaction**, carboxylic acids are alpha brominated with a mixture of bromine and phosphorus tribromide followed by aqueous workup. Reaction occurs via the acyl bromide and its enol.

$$\begin{array}{c} \underset{\underset{H}{|}}{-\overset{\overset{\displaystyle O}{\|}}{C}-\overset{\displaystyle O}{\overset{\|}{C}}-OH} \xrightarrow{\;Br_2,\,PBr_3\;} \xrightarrow{\;H_2O\;} -\overset{\overset{\displaystyle O}{\|}}{C}-\underset{\underset{Br}{|}}{\overset{\overset{\displaystyle O}{\|}}{C}}-OH \end{array}$$

(central scheme with PBr_3 and H_2O arrows)

$$-\overset{\overset{O}{\|}}{\underset{\underset{H}{|}}{C}}-\overset{O}{\overset{\|}{C}}-Br \;\rightleftharpoons\; \left[\,-\overset{OH}{\overset{|}{C}}=\overset{}{C}-Br\,\right] \xrightarrow{\;Br_2\;} -\overset{O}{\overset{\|}{C}}-\underset{\underset{Br}{|}}{\overset{O}{\overset{\|}{C}}}-Br$$

In the **haloform reaction**, methyl ketones react with molecular halogen and excess hydroxide to yield trihalomethyl intermediates that undergo a subsequent nucleophilic acyl substitution. The products are a carboxylate (or a carboxylic acid if protonated during acidic workup) and haloform, CHX_3.

$$R-\overset{O}{\overset{\|}{C}}-CH_3 \xrightarrow[OH^{\ominus}]{X_2} \rightarrow \rightarrow R-\overset{O}{\overset{\|}{C}}-CX_3 \xrightarrow{\;OH^{\ominus}\;} RCO_2^{\ominus} \;+\; CHX_3$$

Alkylation of an alpha position is possible by generating the enolate with a strong base like **lithium diisopropylamide (LDA)**. The enolate is then used as a nucleophile to attack some alkyl halide. The second step is an S_N2 process and thus is limited to unhindered RX.

$$-\overset{O}{\overset{\|}{C}}-\underset{\underset{|}{}}{\overset{\overset{H}{|}}{C}}- \xrightarrow{\;\overset{\oplus}{Li}\;\overset{\ominus}{N}[CH(CH_3)_2]_2\;} \left[\,-\overset{O}{\overset{\|}{C}}-\overset{\ominus}{\overset{|}{C}}- \;\leftrightarrow\; -\overset{\overset{\ominus}{O}}{\overset{|}{C}}=\overset{}{\overset{|}{C}}-\,\right] \xrightarrow{\;R-X\;} -\overset{O}{\overset{\|}{C}}-\underset{\underset{|}{}}{\overset{\overset{R}{|}}{C}}-$$

Topic Test 2: Alpha Substitution Reactions

True/False

1. Cyclohexanone will react with I_2 and aqueous hydroxide to yield iodoform (CHI_3) and a carboxylate.

2. The acid catalyzed halogenation of aldehydes and ketones occurs via enols but the haloform reaction is via enolates.

Multiple Choice

3. Reaction of 3-pentanone with one equivalent of LDA and then iodomethane yields
   a. 3-hexanone.
   b. 3-methoxypentane.
   c. 2-methyl-3-pentanone.
   d. 2,2-dimethyl-3-pentanone.
   e. None of the above

4. If the second step of adding water was omitted from the HVZ procedure, the product would be
   a. a carboxylate and bromoform.
   b. an alpha hydroxy acyl bromide.
   c. an alpha bromo carboxylic acid.
   d. an alpha bromo acyl bromide.
   e. None of the above

Short Answer

Provide unambiguous structural formulas for the missing organic compounds

5.

6.

Topic Test 2: Answers

1. **False.** These conditions and products are suggestive of the haloform reaction; however, that transformation is limited to methyl ketones. Although some alpha iodination on cyclohexanone may take place, there is no reasonable way to make iodoform under these conditions.

2. **True.** Enolates will not form in acid, so any nucleophilicity of the alpha position is due to enol. In base (haloform conditions), the enolate is generated.

3. **c.** 2-methyl-3-pentanone. The LDA gives the enolate that attacks CH_3I (S_N2 reaction), yielding alpha alkylation (methylation) of the original ketone.

4. **d.** an alpha bromo acyl bromide. The products cited in responses a and b make no sense and product c would result if water were included in the scheme as the second step.

5.   6.

TOPIC 3: ALDOL REACTION OF ALDEHYDES AND KETONES

KEY POINTS

✓ *What is an aldol?*

✓ *What is the mechanism for the base-promoted aldol reaction?*

✓ *Under what conditions will an aldol dehydrate (aldol condensation)?*

✓ *What is a mixed aldol?*

✓ *Under what circumstances can a mixed aldol be prepared?*

Recall from Chapter 5 that addition to the carbonyl of an aldehyde or ketone is common.

$$R-\overset{\overset{\displaystyle O}{\|}}{C}-R' \quad \xrightarrow[\text{NuH}]{\overset{\displaystyle Nu^{\ominus}}{} \quad "H^{\oplus}"} \quad R-\overset{\overset{\displaystyle OH}{|}}{\underset{\underset{\displaystyle R'}{|}}{C}}-Nu$$

If the nucleophile is an enolate, the resulting product will be a beta-hydroxy aldehyde or ketone called an **aldol** (i.e., <u>ald</u>ehyde + alc<u>ohol</u>). Aldehydes and ketones bearing alpha hydrogen react in base to give aldol products where the alpha carbon of one reactant molecule is connected to what was the carbonyl of the other. At elevated temperatures, aldol products can spontaneously dehydrate to give α,β-unsaturated aldehydes or ketones. The overall process is then called an **aldol condensation**.

$$R-CH_2-\overset{\overset{\displaystyle O}{\|}}{C}-H \;+\; \underset{\underset{\displaystyle R}{|}}{CH_2}-\overset{\overset{\displaystyle O}{\|}}{C}-H \;\xrightarrow{\text{base}}\; R-CH_2\cdot\overset{\overset{\displaystyle OH}{|}}{C}H-\underset{\underset{\displaystyle R}{|}}{C}H-\overset{\overset{\displaystyle O}{\|}}{C}-H$$

an aldol

(R = H, alkyl, aryl)

$$R-CH_2\cdot CH=\underset{\underset{\displaystyle R}{|}}{C}-\overset{\overset{\displaystyle O}{\|}}{C}-H \xleftarrow[-H_2O]{\Delta}$$

If one of the carbonyl compounds has no alpha hydrogen and is more reactive toward addition of nucleophiles, then a **mixed aldol** (sometimes called **crossed aldol**) is feasible. Unless the above criteria are met, a mixed aldol reaction is impractical because mixtures of products result.

$$\underset{\substack{\text{Formaldehyde}\\ \text{No }\alpha\text{ hydrogen}\\ \text{Reactive toward Nu}^{\ominus}}}{H-\overset{\overset{\displaystyle O}{\|}}{C}-H} \;+\; CH_3-\overset{\overset{\displaystyle O}{\|}}{C}-\underset{\underset{\displaystyle CH_3}{|}}{\overset{\overset{\displaystyle CH_3}{|}}{C}}-CH_3 \;\xrightarrow{OH^{\ominus}}\; \underset{\text{Mixed Aldol}}{\overset{\overset{\displaystyle OH}{|}}{C}H_2-CH-\overset{\overset{\displaystyle O}{\|}}{C}-\underset{\underset{\displaystyle CH_3}{|}}{\overset{\overset{\displaystyle CH_3}{|}}{C}}-CH_3}$$

Intramolecular aldol reactions are observed if both carbonyl components are in the same molecule. Five- or six-membered ring products are especially favored.

Topic 3: Aldol Reaction of Aldehydes and Ketones   145

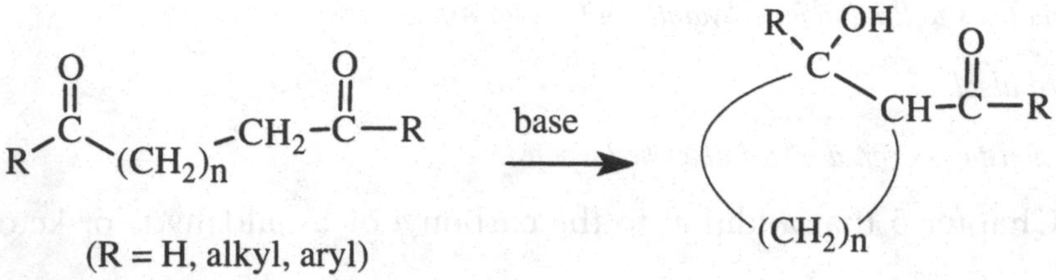

(R = H, alkyl, aryl)

Topic Test 3: Aldol Reaction of Aldehydes and Ketones

True/False

1. The aldol reaction results in the formation of new C—C bonds.
2. If 3-pentanone and acetone are heated with hydroxide in an attempt to carry out a mixed aldol condensation, the resulting reaction will likely yield a mixture of several possible products.

Multiple Choice

3. Which product results from the intramolecular aldol condensation of O═CH(CH$_2$)$_5$CH═O?

   a. HC(=O)CH$_2$CH(OH)CH$_2$CH(OH)CH$_2$C(=O)H

   b. HC(=O)CH=CHCH$_2$CH=CHC(=O)H

   c. 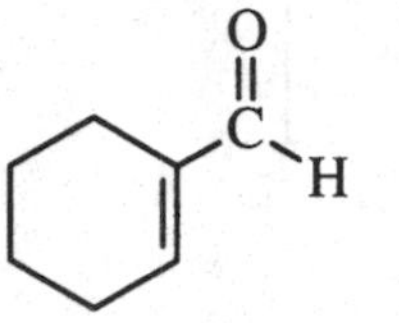

   d. 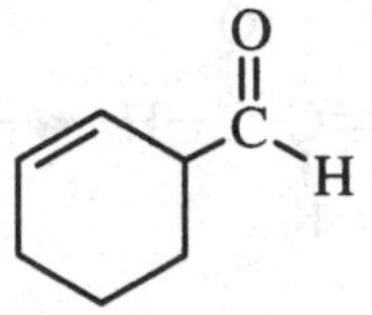

   e. None of these

4. Which is correct concerning an aldol addition reaction?
   a. The product is a beta-hydroxy aldehyde or ketone
   b. The product will dehydrate at higher temperature if possible
   c. At least one reactant had alpha hydrogen
   d. All of the above
   e. None of the above

Short Answer

5. Show the product that results from the mixed aldol addition reaction of acetophenone to formaldehyde (assume no dehydration).
6. Show the structure of the product that would result from the aldol condensation of acetophenone.

Topic Test 3: Answers

1. **True.** The new C—C bond is between the alpha carbon of one aldehyde (or ketone) and what was formerly the C=O carbon or the other.

2. **True.** Because both reactants have alpha hydrogen (and therefore can form enolates) there will likely be two different enolates present. In addition, either enolate can then attack either of the two ketones. After dehydration, the resulting products include 4-ethyl-3-hexen-2-one (from the mixed reaction and with the enolate of acetone), 5-ethyl-4-methyl-4-hepten-3-one (from self-condensation of 3-pentanone), 4-methyl-3-penten-2-one (from self-condensation of acetone), and 4,5-dimethyl-4-hexen-3-one (from the mixed reaction and with the enolate of 3-pentanone).

3. **c.** Note that the alkene is conjugated and resulted from dehydration of the beta-hydroxy aldehyde formed by intramolecular attack of an enolate on the "other" carbonyl.

4. **d.** All of the above

5. [structure: phenyl group attached to C(=O)CH₂CH₂OH]

6. [structure: diphenyl enone]

TOPIC 4: CLAISEN CONDENSATION OF ESTERS

KEY POINTS

✓ *What are the reactants and products of a Claisen condensation?*

✓ *What is the mechanism of the Claisen condensation?*

✓ *What is a mixed Claisen reaction and under what circumstances is it practical?*

✓ *What is a Dieckmann cyclization?*

Recall from Chapter 7 that esters undergo nucleophilic acyl substitution. If the nucleophile is an enolate, the reaction is known as a **Claisen condensation** (also called a **Claisen reaction**). A simple Claisen condensation requires two equivalents of ester. The base used to generate the enolate is normally the same alkoxide that will be the leaving group. The product is a beta-keto ester (i.e., the ketone and ester carbonyl groups are in a 1,3 relationship).

[reaction scheme: RCH₂C(=O)OR′ + CH₂(R)C(=O)OR′ → (Na⁺ OR′⁻ / HOR′, then H₃O⁺) → RCH₂C(=O)—CH(R)C(=O)OR′ (+ HOR′)]

Mixed Claisen reactions are practical in cases where one ester component has no alpha hydrogen and therefore cannot form an enolate. For a mixed Claisen reaction, the enolate can be from an ester or a ketone. In the latter case, the product will be a beta diketone.

The Claisen reaction schemes:

$$\text{C}_6\text{H}_5\text{COOCH}_3 + \text{CH}_3\text{CH}_2\text{COOCH}_3 \xrightarrow{\text{NaOCH}_3} \xrightarrow{\text{H}_3\text{O}^{\oplus}} \text{C}_6\text{H}_5\text{CO-CH(CH}_3\text{)-CO-OCH}_3$$

$$\text{C}_6\text{H}_5\text{COOCH}_3 + \text{(cyclohexanone)} \xrightarrow{\text{NaOCH}_3} \xrightarrow{\text{H}_3\text{O}^{\oplus}} \text{C}_6\text{H}_5\text{CO-(2-oxocyclohexyl)}$$

If both carbonyl components of the Claisen reaction are within the same molecule, the intramolecular process is called the **Dieckmann cyclization** but otherwise is essentially the same as other Claisen reactions. Five- or six-membered ring products are especially favored.

$$\text{ROCCH}_2(\text{CH}_2)_n\text{COR} \xrightarrow[\text{HOR}]{\ominus\text{OR}} \xrightarrow{\text{H}_3\text{O}^{\oplus}} \text{ROC-CH-C (cyclic with (CH}_2)_n)$$

Topic Test 4: Claisen Condensation of Esters

True/False

1. The Claisen condensation is a nucleophilic acyl substitution reaction.

2. An intramolecular aldol condensation is sometimes called a Dieckmann cyclization.

Multiple Choice

3. The product of a Claisen condensation between two esters is usually
   a. an alpha-hydroxy ester.
   b. a beta-hydroxy ester.
   c. an alpha-keto ester.
   d. a beta-keto ester.
   e. None of the above

4. Which product below results from treating ethyl acetate with ethoxide in ethanol followed by aqueous acid workup.

   a. $\text{CH}_3\text{CCH}_2\text{COCH}_2\text{CH}_3$

   b. $\text{CH}_2(\text{CO}_2\text{CH}_2\text{CH}_3)_2$

   c. (2,4-pentanedione structure)

   d. All of these
   e. None of these

Short Answer

Provide unambiguous structural formulas for the missing organic compounds.

5.

$$\text{C}_6\text{H}_5\text{—CH}_2\overset{\overset{\displaystyle O}{||}}{\text{C}}\text{OCH}_2\text{CH}_3 \quad + \quad \text{H}\overset{\overset{\displaystyle O}{||}}{\text{C}}\text{—OCH}_2\text{CH}_3 \quad \xrightarrow[\text{HOCH}_2\text{CH}_3]{\text{NaOCH}_2\text{CH}_3} \quad \xrightarrow{\text{H}_3\text{O}^{\oplus}}$$

6.

$$\text{CH}_3\overset{\overset{\displaystyle O}{||}}{\text{O}}\text{CCH}_2\text{CH}_2\overset{\overset{\displaystyle CH_3}{|}}{\underset{\underset{\displaystyle CH_3}{|}}{\text{C}}}\text{CH}_2\text{CH}_2\overset{\overset{\displaystyle O}{||}}{\text{C}}\text{OCH}_3 \quad \xrightarrow[\text{HOCH}_3]{\text{NaOCH}_3} \quad \xrightarrow{\text{H}_3\text{O}^{\oplus}}$$

Topic Test 4: Answers

1. **True.** The reaction has the same form as most of those in Chapter 7. In this case the nucleophile is the enolate of an ester and the leaving group is an alkoxide (ultimately an alcohol).

2. **False.** That term is used for an intramolecular *Claisen* condensation (not aldol).

3. **d.** a beta-keto ester.

4. **a.** The compounds shown in b and c are a 1,3-diester and a 1,3-diketone, respectively. They could not have resulted from a Claisen condensation of two esters.

5. $\text{H}\overset{\overset{\displaystyle O}{||}}{\text{C}}\text{CH}\overset{\overset{\displaystyle O}{||}}{\text{C}}\text{OCH}_2\text{CH}_3 \quad + \quad \text{HOCH}_2\text{CH}_3$

6.

TOPIC 5: REACTIONS OF STABILIZED ENOLATES

KEY POINTS

✓ *What is important about 1,3-dicarbonyl compounds and their enolates?*

✓ *What are the reactants and products in the malonic ester synthesis?*

✓ *What are the reactants and products in the acetoacetic ester synthesis?*

✓ *What is Michael addition?*

Enolates resulting from proton abstraction from between two carbonyls or other electron-withdrawing groups are resonance stabilized and easily formed. They are well suited for organic synthesis schemes. Enolates from 1,3-dicarbonyl compounds are the most popular, but others are also possible.

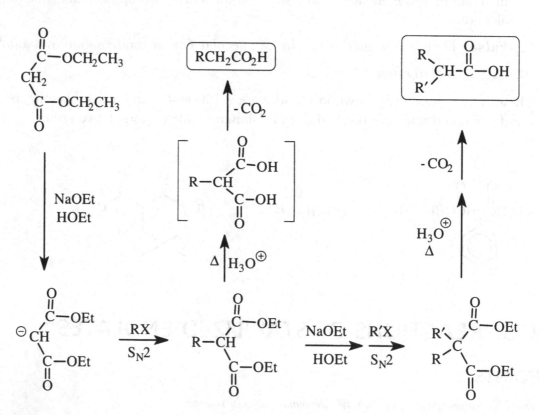

Diethyl malonate (also called malonic ester) can be used to prepare mono- or disubstituted acetic acids in a sequence known as the **malonic ester synthesis**. An enolate is alkylated once (or twice) followed by hydrolysis of the ester groups and thermal decarboxylation. The alkylations occur by S_N2 reactions, thus limiting R and R′ to methyl, primary, and some secondary groups.

(Et = CH_2CH_3; R, R′ ≠ aryl, vinyl, 3° alkyl)

The closely related synthetic scheme, called the **acetoacetic ester synthesis**, provides a route to methyl ketones.

$$CH_3CCH_2COEt \xrightarrow[\text{EtOH}]{\text{NaOEt}} \xrightarrow{RX} \underset{\underset{R}{|}}{CH_3CCHCOEt} \xrightarrow[(-CO_2, -EtOH)]{H_3\overset{\oplus}{O}, \Delta} CH_3C-CH_2-R$$

$$\downarrow \begin{array}{c}\text{NaOEt}\\ \text{EtOH}\end{array}$$

$$\downarrow R'X$$

$$\underset{\underset{R' \quad R'}{C}}{CH_3C \qquad COEt} \xrightarrow[(-CO_2, -EtOH)]{H_3\overset{\oplus}{O}, \Delta} \underset{R}{CH_3C-\overset{R'}{CH}}$$

Recall from Chapter 5 that α,β-unsaturated carbonyl compounds can be attacked at the carbonyl (normal addition) or at the beta carbon (conjugate addition) depending on the nucleophile. Most stabilized enolates are conjugate addition nucleophiles. The conjugate addition of an enolate is called **Michael addition** or the **Michael reaction**. Note that the products are 1,5-dicarbonyl compounds.

Topic Test 5: Reactions of Stabilized Enolates

True/False

1. Acetophenone is a methyl ketone and therefore can be prepared by the acetoacetic ester synthesis.

2. Enolate attack on the carbonyl of an α,β-unsaturated carbonyl compound is called Michael addition.

Multiple Choice

3. Treating acetoacetic ester with ethoxide/ethanol, then benzyl bromide, then hot aqueous acid would yield
   a. 3-phenyl-2-propanone.
   b. 4-phenyl-2-butanone.
   c. 3-phenyl-2-butanone.
   d. 1,3-diphenyl-2-propanone.
   e. None of the above

4. Which compound below could be prepared by the malonic ester synthesis?
   a. Benzoic acid

b. Formic acid

c. 2,2-Dimethylpropanoic acid

d. All of the above

e. None of the above

Short Answer

5. Show how one could prepare 2-methylhexanoic acid by the malonic ester synthesis.

6. Provide unambiguous structural formulas for the missing organic compound(s)

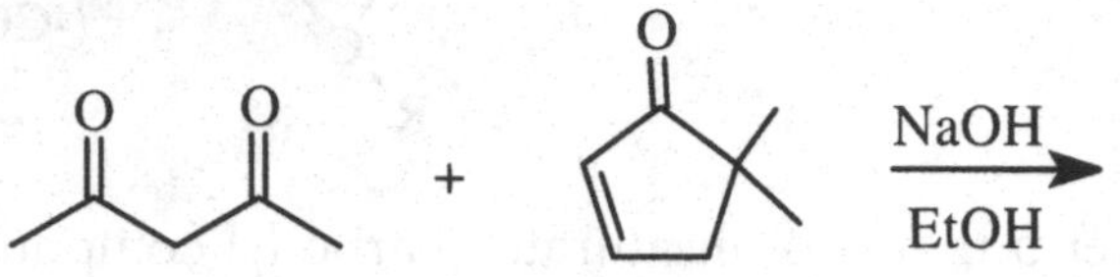

Topic Test 5: Answers

1. **False.** Although it is true that acetophenone is a methyl ketone, it is not of the form RR′CH—CO—CH$_3$.

2. **False.** The enolate must attack the *beta carbon* of the unsaturated carbonyl to lead to Michael addition.

3. **b.** 4-Phenyl-2-butanone, PhCH$_2$—CH$_2$CO—CH$_3$. The benzyl group (PHCH$_2$—) easily placed on the central alpha carbon of the enolate and then the ester undergoes hydrolysis and decarboxylation.

4. **e.** None of the above. None of the compounds listed can be viewed as mono- or disubstituted acetic acids (RCH$_2$CO$_2$H or RR′CHCO$_2$H, respectively).

5.

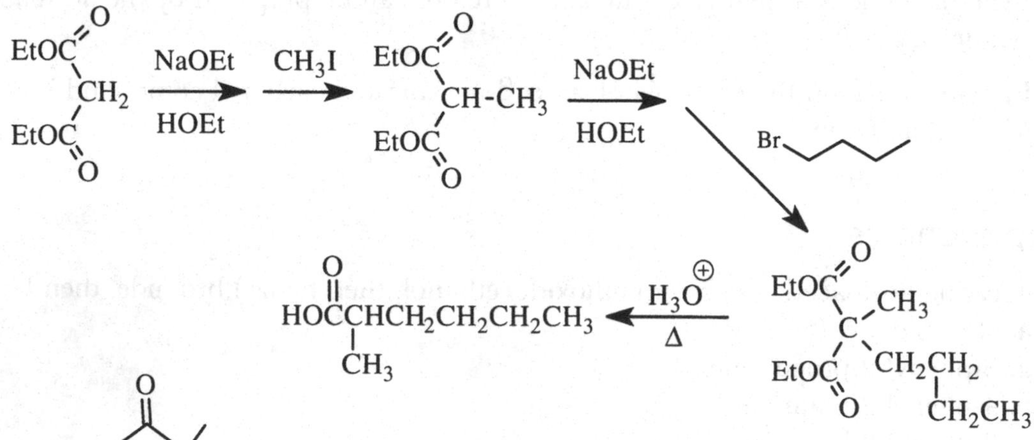

6.

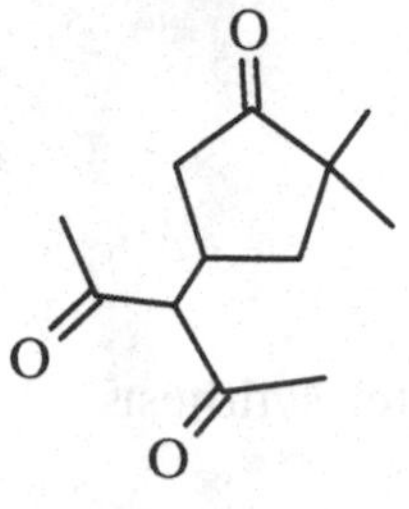

TOPIC 6: MORE COMPLEX REACTIONS OF STABILIZED ENOLATES

KEY POINTS

✓ *What is the general form of reactants and products for a Robinson annulation?*

✓ *What is the mechanism of a Robinson annulation?*

✓ *What is the general form of reactants and products for the Stork enamine synthesis?*

✓ *What is the mechanism of the Stork enamine synthesis?*

The **Robinson annulation** is a strategy for preparing conjugated cyclohexenones (i.e., the product is a new α,β-unsaturated six-membered ring ketone). The enolate is generated in the presence of an α,β-unsaturated carbonyl compound. A Michael reaction followed by an intramolecular aldol reaction and dehydration gives the product in a one-pot process.

Best results for Michael reactions are obtained if the enolate is from a 1,3-dicarbonyl compound. One can achieve the same net transformation from a simple monoketone or monoaldehyde by first converting it to an enamine. Recall from Chapter 5 that aldehydes and ketones bearing alpha hydrogen react with secondary amines to give enamines.

Enamines are conjugate addition nucleophiles. After addition to an α,β-unsaturated carbonyl compound, the original carbonyl can be regenerated by hydrolysis. These steps are combined in the **Stork enamine synthesis** that, in effect, provides the path to products formally arising from Michael addition of simple enolates. The products will be 1,5-dicarbonyl compounds.

If the electrophile is a simple alkyl or acyl halide, enamines can also be used as enolate equivalents in alkylation or acylation reactions.

Topic Test 6: More Complex Reactions of Stabilized Enolates

True/False

1. The Robinson annulation is a combination of the Michael reaction and an intramolecular aldol condensation.

2. Enamines are nucleophilic at the beta carbon.

Multiple Choice

3. Robinson annulation products are
   a. six-membered rings.
   b. conjugated.
   c. ketones.
   d. All of the above
   e. None of the above

Short Answer

Provide unambiguous structural formulas for the missing organic compounds.

4. 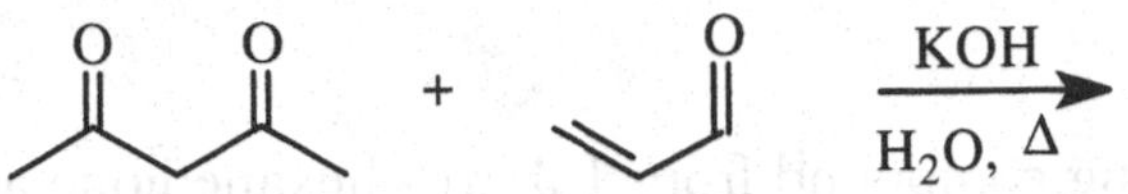 + $\xrightarrow[\text{H}_2\text{O, } \Delta]{\text{KOH}}$

5.

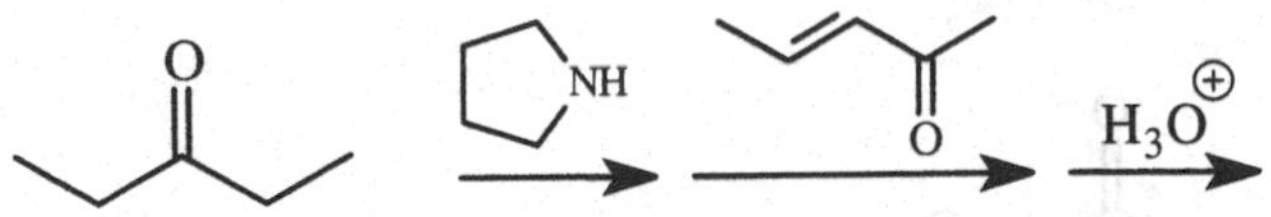

6. Show how one could convert acetone into 2-butanone via an enamine synthesis.

Topic Test 6: Answers

1. **True**

2. **True.** Consider resonance delocalization of the lone electron pair of nitrogen into the C=C bond.

3. **d.** All of the above

4. 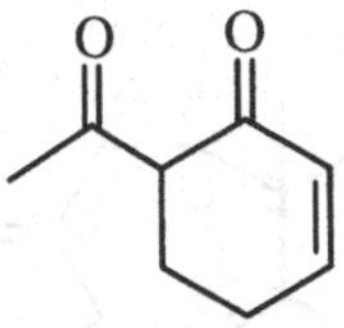   5.

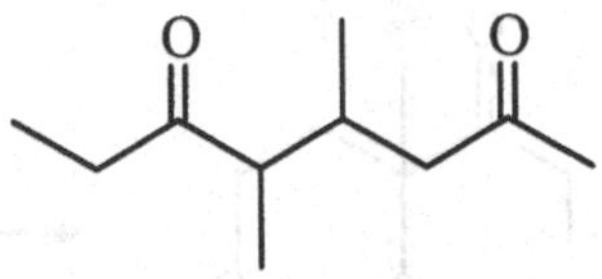

6.

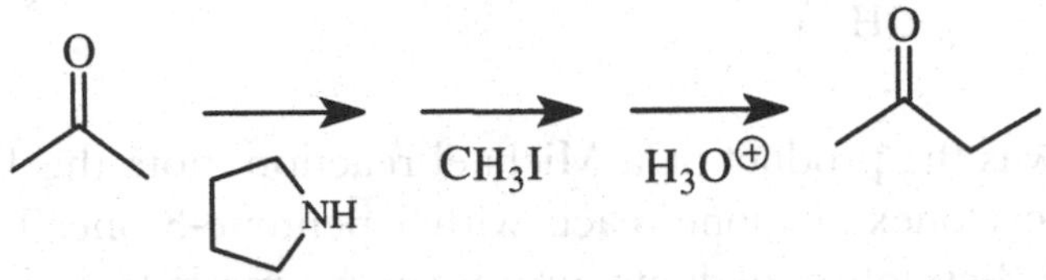

APPLICATION

Examples of aldol-like and Claisen-like reactions are abundant in biochemistry. During periods of fasting or starvation, the brain's required supply of glucose comes from a series of biochemical transformations called gluconeogenesis (glucose + new + creation). A key step in gluconeogenesis formally results from a mixed aldol reaction of two different three-carbon reactants. The enzyme aldolase, rather than base, catalyzes this reversible reaction. The product is a six-carbon beta-hydroxy ketone. (The new C—C bond is shown as a broken line for emphasis.)

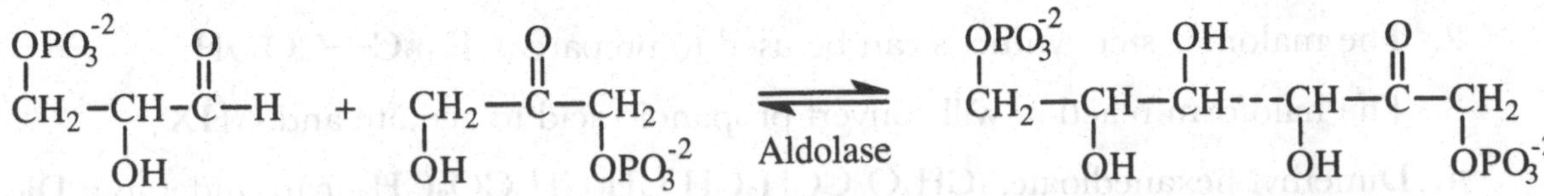

DEMONSTRATION PROBLEM

Show how one could synthesize the following compound from 1,3-cyclohexanedione and any other needed reagents. More than one step is required.

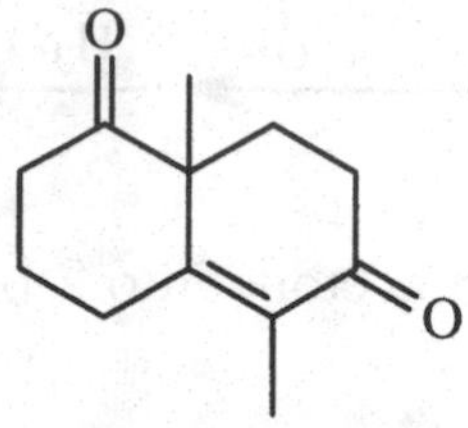

Solution

As always, the way to analyze this synthesis problem is by thinking backward from the desired product. Note that the product has two fused six-membered rings, one of which is an α,β-unsaturated cyclohexenone. This suggests it came from a Robinson annulation. (The other ring was present in the starting material.) Recall the last step(s) of a Robinson annulation is dehydration of the beta-hydroxy ketone that resulted from an intramolecular aldol reaction.

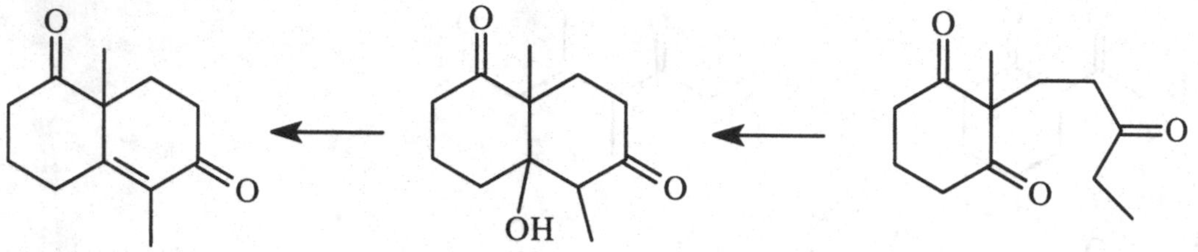

The precursor to the aldol reaction was the product of a Michael reaction (note the 1,5 dicarbonyl). The enolate of 2-methyl-1,3-cyclohexanedione reacts with 1-pentene-3-one. The starting material specified in the problem is 1,3-cyclohexanedione, which can be methylated in two steps. The entire synthesis sequence is shown below.

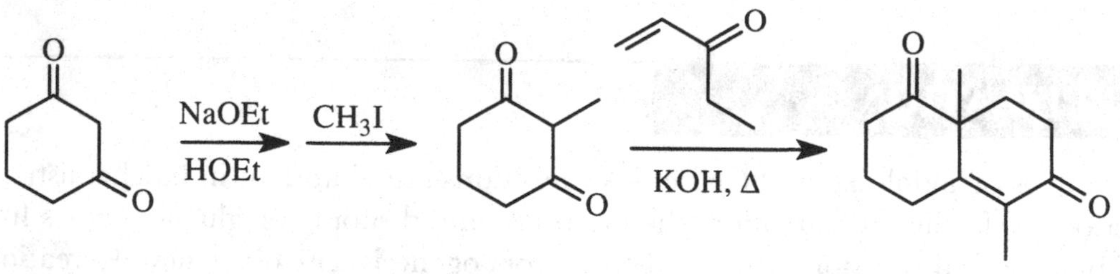

Chapter Test
True/False

1. The Claisen condensation of ethyl acetate yields acetoacetic ester.

2. The malonic ester synthesis can be used to prepare $(CH_3)_3C$—COOH.

3. The haloform reaction will convert propanoic acid to acetate and CHX_3.

4. Dimethyl hexanedioate, $(CH_3O_2CCH_2CH_2CH_2CH_2CO_2CH_3)$ can undergo a Dieckmann cyclization to yield a product with a six-membered ring.

Multiple Choice

5. The HVZ reaction
   a. works only on methyl ketones.
   b. involves an enamine intermediate.
   c. produces an alpha-bromocarboxylic acid.
   d. All of the above
   e. None of the above

6. Which reagent below would be useful if preparing pentanoic acid from malonic ester?
   a. 1-Bromopropane
   b. 2-Bromopropane
   c. 1-Bromopentane
   d. 2-Bromopentane
   e. None of the above

7. An intramolecular aldol condensation
   a. is called a Dieckmann cyclization.
   b. is called a Michael addition.
   c. is part of a Robinson annulation.
   d. produces a beta-keto ester.
   e. None of the above

8. Which product results if acetaldehyde ($CH_3CH{=}O$) is converted to an enamine that is then reacted with $CH_2{=}CH{-}CH{=}O$ followed by aqueous acid hydrolysis?
   a. 2-Cyclohexenone
   b. 3-Hydroxy-4-pentenal
   c. Malonic ester
   d. $O{=}CHCH_2CH_2CH_2CH{=}O$
   e. None of the above

9. The acetoacetic ester synthesis is used to prepare
   a. α,β-unsaturated carbonyl compounds.
   b. methyl ketones.
   c. enamines.
   d. benzoic acid.
   e. None of the above

What product results from reacting cyclohexanone under each set of conditions indicated?

10. Br_2, CH_3CO_2H

11. LDA, then $CH_3CH_2CH_2Br$

12. KOH, H_2O, heat

13. Draw all reasonable resonance forms of the enolate that results from treating 2-ethyl-1,3-cyclohexanedione with one equivalent of base.

14.

15. 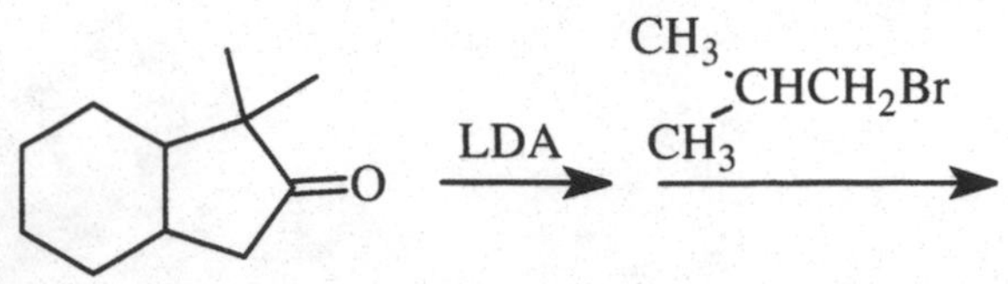 LDA, CH₃CHCH₂Br/CH₃ →

16. 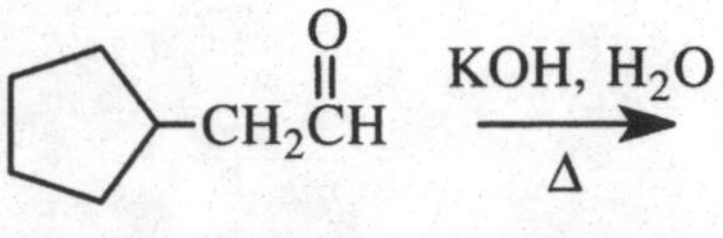 KOH, H₂O, Δ →

17. 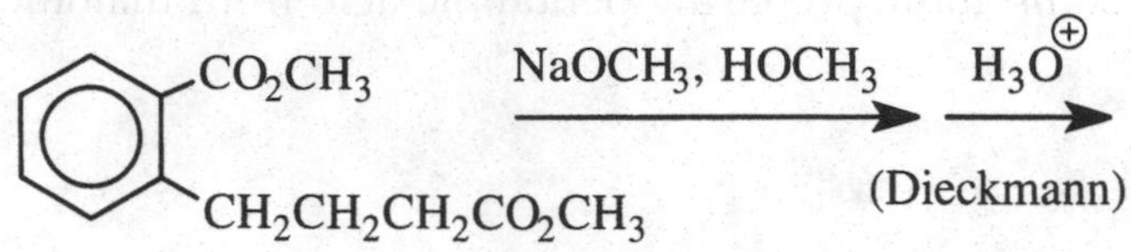 NaOCH₃, HOCH₃ → H₃O⁺ → (Dieckmann)

18.

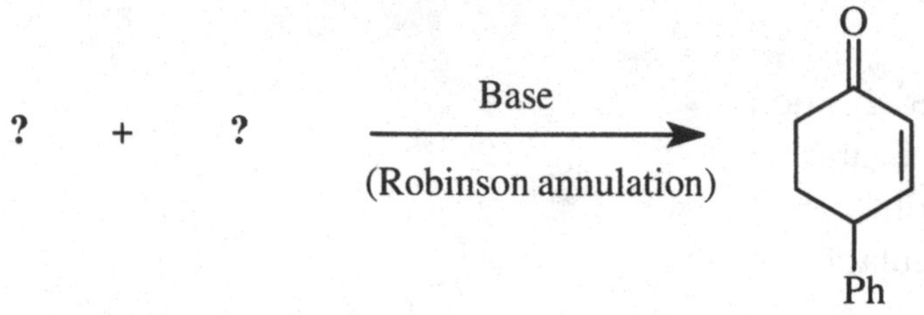

? + ? →(Base, Robinson annulation)→

19. 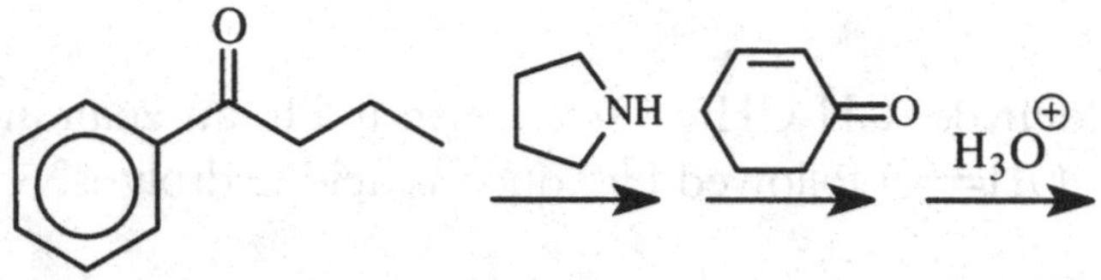 →(pyrrolidine NH)→ →(cyclohexenone)→ H₃O⁺ →

20. 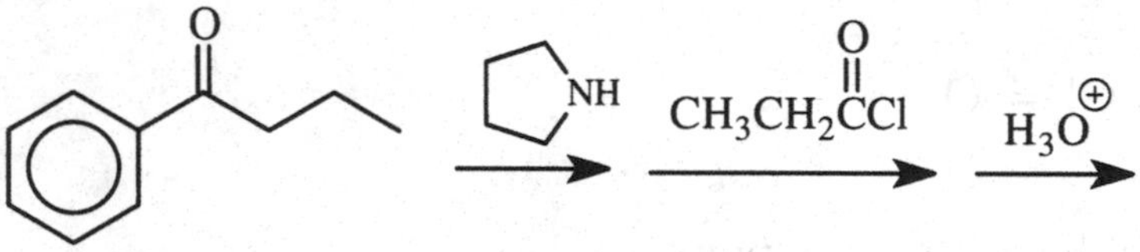 →(pyrrolidine NH)→ CH₃CH₂CCl → H₃O⁺ →

21. 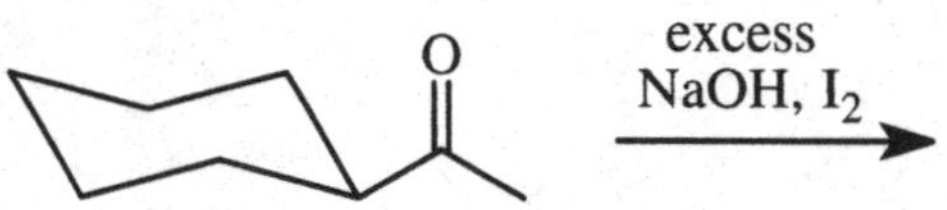 excess NaOH, I₂ →

22. Show how to make 3-ethyl-2-pentanone from acetoacetic ester.

Chapter Test: Answers

1. **True**

2. **False**

3. **False**

4. **False**

5. **c** 6. **a** 7. **c** 8. **d** 9. **b**

10. 2-Bromocyclohexanone

11. 2-Propylcyclohexanone

12.

13.

14.

15.

16.

17.

18.

19.

20.

21. $CO_2^\ominus$ + CHI_3

22. $CH_3CCH_2COCH_2CH_3$ $\xrightarrow[\text{HOEt}]{\text{NaOEt}}$ $\xrightarrow{CH_3CH_2Br}$ $\xrightarrow[\text{HOEt}]{\text{NaOEt}}$ $\xrightarrow{CH_3CH_2Br}$

CH_3CCH with CH_2CH_3 and CH_2CH_3 groups $\xleftarrow{H_3O^\oplus}$

Check Your Performance

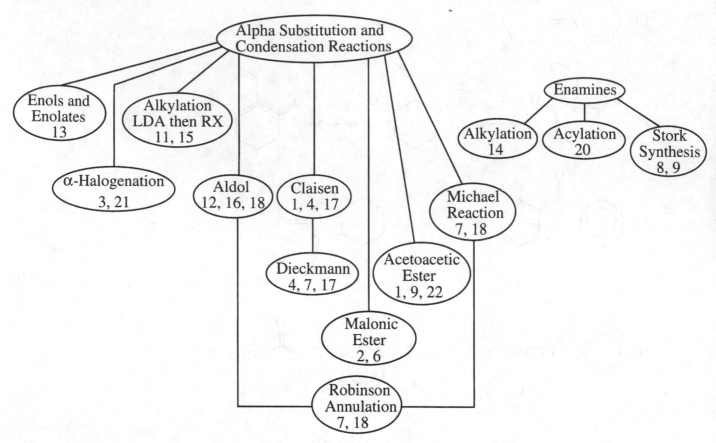

Note the number of questions in each grouping that you got wrong on the chapter test. Identify areas where you need further review and go back to relevant parts of this chapter.

CHAPTER 9

Amines

Some organic compounds isolated from plants are basic in aqueous solution and were once called "vegetable alkali." More recently these have become known as alkaloids. Many alkaloids are well known, as shown in the examples below. The structural feature responsible for the basicity of alkaloids is the amine functional group. In this chapter we survey the chemistry of amines.

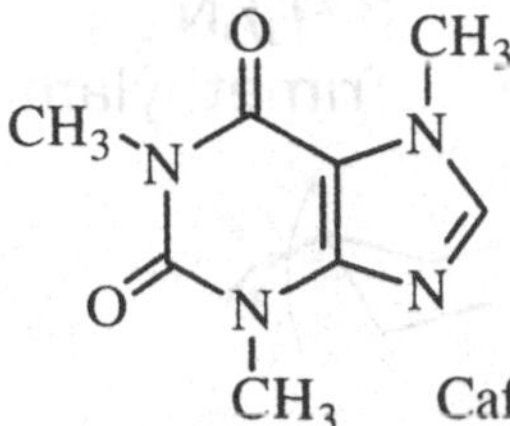

Caffeine
(Coffee, Tea)

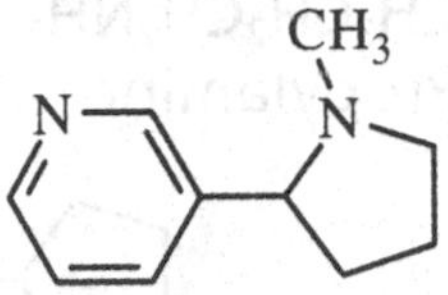

Nicotine
(Tobacco)

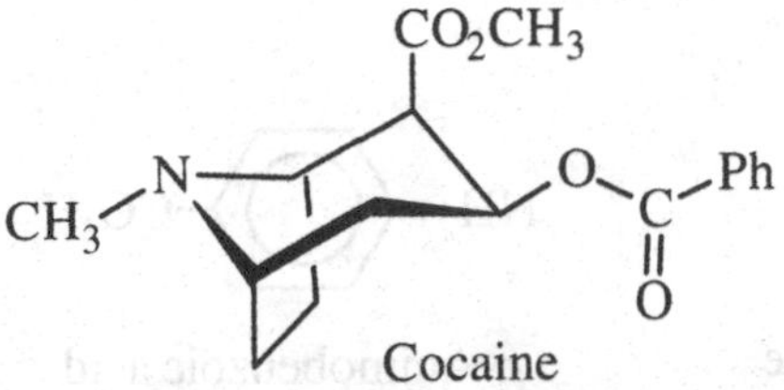

Cocaine
(Coca leaves)

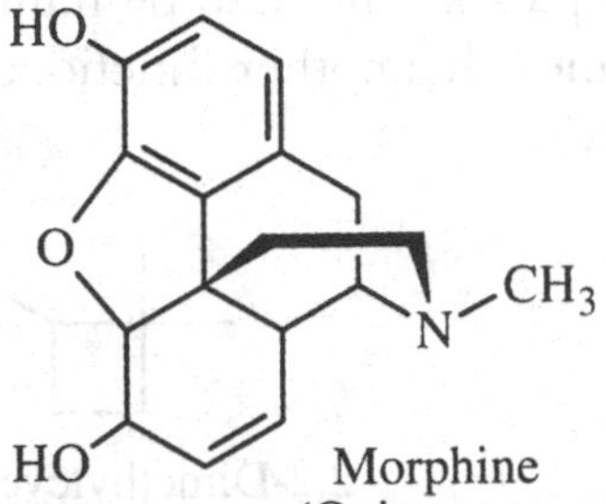

Morphine
(Opium poppy)

ESSENTIAL BACKGROUND

- **Acid-base reactions**
- **Nucleophilic substitution on alkyl halides by S_N2**
- **Elimination reactions and Zaitsev's rule**
- **Electrophilic aromatic substitution (EAS) (Chapter 1)**
- **Electron-donating and -releasing substituents (Chapter 1)**
- **Nucleophilic aromatic substitution (Chapter 1)**
- **Imine formation from aldehydes and ketones (Chapter 5)**
- **Reduction of amides and nitriles with lithium aluminum hydride (Chapter 7)**
- **Aminolysis of acyl halides, anhydrides, or esters (Chapter 7)**
- **Malonic ester synthesis (Chapter 8)**

Topic 1: Structure, Classification, Nomenclature, and Properties   161

TOPIC 1: STRUCTURE, CLASSIFICATION, NOMENCLATURE, AND PROPERTIES

KEY POINTS

✓ *How are amines classified as primary, secondary, tertiary, etc.?*

✓ *How are amines and related compounds named?*

✓ *What are the most important chemical and physical properties of amines?*

✓ *How can one predict the relative basicity of amines?*

Amines are classified according to the number of carbons attached directly to nitrogen. Primary, secondary, and tertiary amines have one, two, or three alkyl carbons replacing the hydrogen atoms of ammonia. Simple primary amines and symmetrical secondary or tertiary amines are named by identifying the alkyl group(s) and adding the suffix "amine."

$CH_3CH_2CH_2NH_2$ $(CH_3CH_2)_2NH$ $(CH_3)_3N$
Propylamine   Diethylamine   Trimethylamine

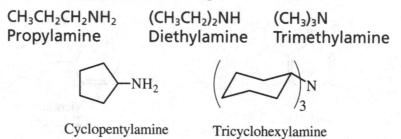

Cyclopentylamine   Tricyclohexylamine

More complex amines can be named by replacing the final "e" of the parent alkane name with "amine." If another functional group is present, the NH_2 group is considered an amino substituent.

2,2-Dimethylcyclobutaneamine   *p*-Aminobenzoic acid

Unsymmetrical secondary and tertiary amines are named as *N*-substituted primary amine derivatives.

$CH_3CH_2CH_2CH_2CH_2CH_2N$ with CH_3 and CH_2CH_3

N-Ethyl-*N*-methylhexylamine

N,N-Diethylcycloheptylamine

Arylamines contain a nitrogen atom bound directly to an aromatic ring. These include aniline and its derivatives.

NH_2—◯—NO_2 ◯—$NHCH_3$

p-Nitroaniline   *N*-Methylaniline

Tetracoordinate nitrogen cations are ammonium ions. Organic derivatives of ammonium can be primary, secondary, tertiary, or quaternary depending upon how many carbons have replaced hydrogen in the parent ammonium ion, NH_4^+.

RNH_3^+
Alkylammonium
Primary

$R_2NH_2^+$
Dialkylammonium
Secondary

R_3NH^+
Trialkylammonium
Tertiary

R_4N^+
Tetraalkylammonium
Quaternary

Names of these ions are similar to those of amines except the suffix "amine" is replaced by "ammonium." Salt names follow the usual pattern with the cation named first then the anion.

$(CH_3CH_2CH_2CH_2)_4N^+F^-$
Tetrabutylammonium
Fluoride

$[CH_3NH_4^+]NO_3^-$
Methylammonium
Nitrate

$[PhCH_2N(CH_3)_3]^+Br^-$
N,N,N-Trimethylbenzylammonium
Bromide

Primary and secondary amines contain N—H bonds and can therefore hydrogen bond. This explains why their boiling points and melting points are higher than those of comparably sized alkanes. Low-molecular-weight amines are at least partially water soluble. The lone pair of electrons on an amine nitrogen makes it a base and a nucleophile. The basicity of alkyl amines is less than that of hydroxide but greater than that of arylamines or amides due to resonance. The relative basicity among a series of alkyl or arylamines is predicted on the basis of electron-donating or electron-withdrawing groups.

$$OH^{\ominus} \quad > \quad RNH_2 \quad > \quad Ar-NH_2 \quad > \quad R-\overset{\overset{\textstyle O}{\|}}{C}-NH_2$$

← Increased Base Strength

NH₂ (EDG) > NH₂ > NH₂ (EWG)

← Increased Base Strength

EDG = Electron Donating Group, EWG = Electron Withdrawing Group

Amines generally react with acids to yield salts. At room temperature, even carboxylic acids react with amines in acid base reactions (rather than nucleophilic acyl substitution to amides). Such ammonium carboxylate salts can be dehydrated at higher temperatures.

$$RCO_2H + HNR_2 \xrightarrow{R.T.} [RCO_2]^-[H_2NR_2]^+ \xrightarrow[-H_2O]{\Delta} \text{Amide}$$

Topic Test 1: Structure, Classification, Nomenclature, and Properties

True/False

1. Trimethylamine has a higher boiling point than the isomer propylamine due to hydrogen bonding.

2. *tert*-Butylamine, $(CH_3)_3C$—NH_2, is classified as a tertiary amine.

Multiple Choice

3. Which statement applies to the species $^+N(CH_3)_4$?
   a. It is called butyl ammonium.
   b. It is capable of hydrogen bonding.
   c. It reacts with acids in acid-base reactions.
   d. All of the above
   e. None of the above

4. Which of the following is the strongest base?

   a. b. c.

   d. e.

Short Answer

5. Provide an unambiguous structural formula for *N*-butylcyclohexylamine.

6. Name the following compound.

Topic Test 1: Answers

1. **False.** Hydrogen bonding is not possible for trimethylamine because it is a tertiary amine and has no N—H bonds. It will therefore have weaker intermolecular attractions and a lower boiling point than will propylamine (a primary amine).

2. **False.** Although *tert*-butyl is a tertiary alkyl group, the nitrogen atom of the amine is connected to only one carbon so it is classified as a primary amine.

3. **e.** None of the above. The correct name for this ion is tetramethylammonium, and because it is a quaternary ammonium ion it does not have any N—H bonds or an unshared pair of electrons on nitrogen. Those structural deficiencies forbid hydrogen bonding or protonation by an acid.

4. **b.** $PhCH_2NH_2$. Benzylamine is the only compound among those shown in which the lone pair of electrons on nitrogen is not partially delocalized by resonance (a, c, and d) or already protonated and therefore incapable of reacting as a base (e). Basicity requires the electron pair to be available to form a bond to a proton.

5. 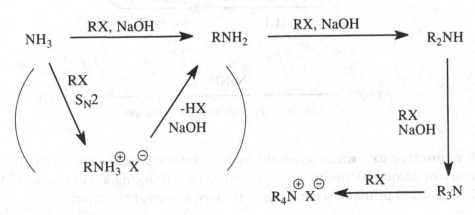 NHCH$_2$CH$_2$CH$_2$CH$_3$

6. Dicyclopropylamine

TOPIC 2: PREPARATION OF AMINES

KEY POINTS

✓ *What reactions from previous chapters will produce amines?*

✓ *How are amines prepared from alkyl halides?*

✓ *What reagents and steps are used in the Gabriel synthesis?*

✓ *What are the reactants and products of reductive amination?*

✓ *What are the reactants and products of the Hofmann and Curtius rearrangements?*

✓ *How can arylamines be prepared from aromatic nitro compounds?*

Recall from Chapter 8 that amides can be reduced with lithium aluminum hydride to yield primary, secondary, or tertiary amines but at least one of the alkyl groups must be primary.

$$R-\overset{\overset{\displaystyle O}{\|}}{C}-N\overset{\nearrow R'}{\underset{\searrow R''}{}} \xrightarrow[\text{ether}]{LiAlH_4} \xrightarrow{H_2O} R-CH_2-N\overset{\nearrow R'}{\underset{\searrow R''}{}}$$

$$(R, R', R'' = H, \text{alkyl}, \text{aryl})$$

Under similar conditions, a nitrile will lead to a primary amine bearing a primary alkyl group.

$$RC\equiv N \xrightarrow[\text{ether}]{LiAlH_4} \xrightarrow{H_2O} RCH_2NH_2$$

Although ammonia and amines react with alkyl halides by S_N2 to yield ammonium salts that in the presence of excess amine or some added base like NaOH will give amine products, multiple substitution often leads to mixtures, making this method impractical for synthesis of most amines.

Better results are obtained with other nitrogen nucleophiles. **Azide ion** can displace a halide by S_N2. The resulting alkyl azide is then reduced of with lithium aluminum hydride. This method is limited to preparation of primary amines.

$$R-X \xrightarrow{\overset{\oplus}{Na} \overset{\ominus}{N_3}} R-N_3 \xrightarrow{LiAlH_4} \xrightarrow{H_3O^{\oplus}} R-NH_2$$

$$(R-\overset{\oplus}{N}=N=\overset{\ominus}{N})$$

In the **Gabriel amine synthesis**, the starting material, phthalimide, is deprotonated to the phthalimide anion. That nucleophile reacts with alkyl halides by S_N2 to produce *N*-alkylated imides that, in turn, undergo alkaline hydrolysis to yield primary amines and phthalate.

Arylamines (aniline and its derivatives) can be prepared by reduction of the corresponding nitro compounds or by nucleophilic aromatic substitution (Chapter 1) on aryl halides with the NH_2^- nucleophile.

Through **reductive amination**, an aldehyde or ketone is converted to an amine. Ammonia or a primary or secondary amine in the presence of a reducing agent (such as H_2/Ni or $NaBH_3CN$) leads to primary, secondary, or tertiary amines, respectively.

The reaction scheme at top:

$$R-\overset{O}{\overset{\|}{C}}-R' \xrightarrow[\text{NaBH}_3\text{CN} \text{ or H}_2, \text{Ni}]{\underset{R'''}{\overset{R''}{HN}}} \underset{R-CH-R'}{\overset{R''-N-R'''}{}} \quad (R, R', R'', R''' = H, alkyl, aryl)$$

Heating an acyl azide leads to a loss of N_2, yielding an isocyanate that will undergo hydrolysis and decarboxylation to a primary amine. The process is called the **Curtius rearrangement**. The acyl azide starting material is normally prepared from the corresponding acyl chloride.

$$R-\overset{O}{\overset{\|}{C}}-Cl \xrightarrow{N_3^{\ominus}} R-\overset{O}{\overset{\|}{C}}-N_3 \xrightarrow{\Delta} R-N=C=O$$
$$\text{Acyl azide} \qquad \text{Isocyanate}$$

$$RNH_2 \xleftarrow{-CO_2} R-NH-\overset{O}{\overset{\|}{C}}-OH \xleftarrow{H_2O}$$

The mechanistically similar **Hofmann rearrangement** converts a primary amide to a primary amine.

$$R-\overset{O}{\overset{\|}{C}}-NH_2 \xrightarrow[\text{H}_2\text{O}]{X_2, \text{NaOH}} RNH_2 + CO_2$$

$$\downarrow$$
$$\downarrow$$

$$R-\overset{O}{\overset{\|}{C}}-NHX \longrightarrow R-N=C=O$$

Topic Test 2: Preparation of Amines

True/False

1. None of the carbons of the phthalimide starting material are found in the amine product made by a Gabriel amine synthesis.

2. If acetyl chloride is treated with sodium azide and the resulting product is heated and then treated with water, a Curtius rearrangement will yield methylamine.

Multiple Choice

3. Which method below can be used to prepare a tertiary amine?
   a. Reduction of an amide with LiAlH$_4$ then water
   b. The Gabriel amine synthesis
   c. The Hofmann rearrangement
   d. All of the above
   e. None of the above

4. Which reaction below is best suited for the synthesis of aniline?
   a. Reductive amination of benzaldehyde with ammonia and $NaBH_3CN$
   b. Treatment of bromobenzene with NaN_3 then $LiAlH_4$ followed by water
   c. Treatment of benzamide ($PHCONH_2$) with Br_2 and hot aqueous $NaOH$
   d. All of the above
   e. None of the above

Short Answer

5. How one could make *N*-ethylbenzylamine by reductive amination.

6. How could one prepare *p*-toluidine (*p*-methylaniline or *p*-aminotoluene) from toluene. More than one step may be required.

Topic Test 2: Answers

1. **True.** Only the phthalimide nitrogen makes it into the amine product. All the carbons end up in the phthalate ion byproduct.

2. **True.** $CH_3COCl \rightarrow CH_3CON_3 \rightarrow CH_3-N{=}C{=}O \rightarrow CH_3NH_2 + CO_2$

3. **a.** Reduction of an amide with $LiAlH_4$ then water. A tertiary amide will reduce under these conditions to yield a tertiary amine. The Gabriel amine synthesis and the Hofmann rearrangement are both limited to the preparation of primary amines.

4. **c.** Treatment of benzamide ($PHCONH_2$) with Br_2 and hot aqueous $NaOH$. These are the Hofmann rearrangement conditions. Choice a would yield benzyl amine, $PhCH_2NH_2$, and choice b would result in no reaction because azide ion cannot displace bromine from an aromatic ring in the first step.

5. Treat benzaldehyde with ethylamine in the presence of either H_2/Ni or $NaBH_3CN$. Alternatively, one could treat acetaldehyde with benzylamine in the presence of either H_2/Ni or $NaBH_3CN$.

6. There is more than one possible correct answer, but the most straightforward method is nitration to *p*-nitrotoluene followed by reduction of the nitro group.

TOPIC 3: REACTIONS OF AMINES

KEY POINTS

✓ *What reactions of amines were covered earlier?*

✓ *What are the reagents, mechanism, and regiochemistry of Hofmann elimination?*

✓ *What are aryldiazonium compounds and how are they prepared from amines?*

✓ *What are the reactants and products of Sandmeyer reactions and related reactions?*

✓ *How are aryldiazonium compounds used in EAS reactions?*

Some reactions of amines have already been discussed. We saw in Topics 1 and 2 that amines can be protonated in acid-base reactions and can also be alkylated by S_N2 reactions. Recall that amines react with acyl halides, anhydrides, or esters to yield amides (aminolysis, Chapter 7). Aniline and its derivatives are activated for EAS reactions (Chapter 1).

$$NH_2R \xrightarrow{HA} [RNH_3^{\oplus}]\,A^{\ominus}$$

$$NH_2R \xrightarrow[NaOH]{R'X} NHRR' \ (+ \text{ others})$$

$$NH_2R \xrightarrow{R'-\overset{O}{\overset{\|}{C}}-L} R'-\overset{O}{\overset{\|}{C}}-NHR$$

$$(L = X, O_2CR'', \text{ or } OR')$$

If an amine is exhaustively methylated with excess iodomethane and then heated with silver oxide, the net transformation is elimination of an amine to an alkene. This is called **Hofmann elimination**. The mechanism is E2 in which the leaving group is a tertiary amine with at least one methyl group. The regiochemistry is **non-Zaitsev** (i.e., the least substituted alkene is favored).

$$CH_3CH_2CH_2CH=CH_2 \ + \ CH_3CH_2CH=CHCH_3$$
$$\text{(major)} \qquad\qquad\qquad \text{(minor)}$$

$$+ \ N(CH_3)_3$$

When primary arylamines are treated with nitrous acid HONO, a reaction known as **diazotization** occurs to yield a moderately stable **aryldiazonium** ion, ArN_2^+.

$$Ar-NH_2 \xrightarrow[H_2SO_4]{HONO} [ArN_2^{\oplus}][HSO_4^{\ominus}] \quad \text{an aryldiazonium salt}$$

Salts of aryldiazoniums can be converted to a variety of substituted aromatics. Aryl chlorides, bromides, and cyanides are produced by reaction with the copper (I) salts in a process called the **Sandmeyer reaction.**

$$ArN_2^{\oplus} \xrightarrow{\text{CuCl}} ArCl \qquad (+\ N_2)$$

$$ArN_2^{\oplus} \xrightarrow{\text{CuBr}} ArBr$$

$$\xrightarrow{\text{CuCN}} Ar{-}C{\equiv}N$$

Aryldiazonium salts can also be converted to aryl iodides, phenols, or Ar—H by reaction with sodium iodide, aqueous acid, or hypophosphorous acid, respectively.

$$ArN_2^{\oplus} \xrightarrow{\text{NaI}} ArI \qquad (+\ N_2)$$

$$ArN_2^{\oplus} \xrightarrow{H_3O^{\oplus}} ArOH$$

$$\xrightarrow{H_3PO_2} Ar{-}H$$

Aryldiazonium ions can react as electrophiles and couple to activated aromatic rings by EAS reactions to yield azo compounds.

(Y = OH, OR, NH_2, NR_2, etc.)

Topic Test 3: Reactions of Amines

True/False

1. Hofmann elimination obeys Zaitsev's rule and yields the more substituted alkene product where possible.

2. An aryldiazonium will couple to nitrobenzene via an EAS reaction to yield an azo compound.

Multiple Choice

3. Which product results if the reagents below are combined at room temperature?
$(CH_3CH_2)_2NH + CH_3CO_2H \rightarrow$
   a. *N,N*-diethylacetamide
   b. Diethylammonium acetate
   c. Triethylamine
   d. $CH_3CH_2{-}N{=}N{-}CH_2CH_3$
   e. None of the above

4. Aniline could be converted to benzene through which reaction sequence?
   a. H_2SO_4, HNO_3, followed by H_2, Ni
   b. Excess CH_3I followed by aqueous AgO_2 and heat
   c. HONO, H_2SO_4 followed by H_3PO_2

d. All of the above

e. None of the above

Short Answer

5. Show the reagents and conditions one could use to convert 2-ethylaniline into 2-ethylbezenenitrile.

Provide an unambiguous structural formula for the missing organic compound.

6.
excess CH_3I → Ag_2O Δ →

Topic Test 3: Answers

1. **False.** The observed regiochemistry leads normally to less substituted alkenes (non-Zaitsev).

2. **False.** Although aryldiazoniums do couple via EAS reactions on some aromatic rings, the reaction is only successful if the ring on which the EAS reaction is supposed to occur is activated. This normally means there must be a strong activating group on the ring such as NH_2, NHR, NR_2, OH, or OR, each of which bear a nonbonded electron pair in conjugation with the ring. Electron-withdrawing (deactivating) groups like NO_2 render a ring unreactive toward EAS with an aryldiazonium electrophile.

3. **b.** Diethylammonium acetate. This is an acid-base reaction yielding the salt $[(CH_3CH_2)_2NH_2^+][CH_3COO^-]$. Do not be fooled by response a, which is an amide and appears to have resulted from a nucleophilic acyl substitution (Chapter 7) on a carboxylic acid. Recall from Chapter 6 that carboxylic acids react with bases faster than they undergo substitution at the carbonyl. The acid-base reaction in this problem is like those from Chapter 6 but is now presented in the context of the amine base rather than the carboxylic acid.

4. **c.** HONO, H_2SO_4 followed by H_3PO_2. The aniline undergoes diazotization and then the phenyldiazonium is converted to benzene with hypophosphorous acid.
$Ph—NH_2 \rightarrow Ph—N_2^+ \rightarrow Ph—H$

5. There may be more than one possible correct answer; however, the simplest strategy is diazotization followed by a Sandmeyer reaction with sodium cyanide.

$$Ar—NH_2 \xrightarrow{HONO} Ar—N_2^+ \xrightarrow{NaCN} Ar—C\equiv N$$

6. The product 5-(dimethylamino)-1-hexene results from the Hofmann elimination, which gives non-Zaitsev regiochemistry. Because the starting amine is part of a ring, the amine "leaving group" is still tethered to the product.

$$\left[\begin{array}{c} \overset{\oplus}{N(CH_3)_2} \\ CH_3 \end{array} \right] \overset{\ominus}{OH} \longrightarrow \begin{array}{c} N(CH_3)_2 \\ CH_3 \end{array} \equiv CH_3CH_2CH_2CH_2CHCH_3 \\ | \\ N(CH_3)_2$$

The high degree of conjugation in azo compounds resulting from aryldiazonium coupling reactions allows them to absorb energy in the visible region of the electromagnetic spectrum (Chapter 2). That is just a way of saying these compounds have rich vivid colors. These products are often called **azo dyes**. They have been used as food colorings, textile dyes, and pH indicators. As one might expect, the extent and shape of conjugation determines the color observed. Some examples are shown below.

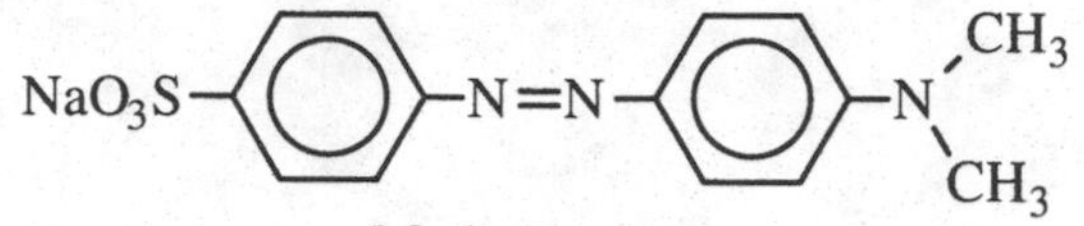

Methyl orange

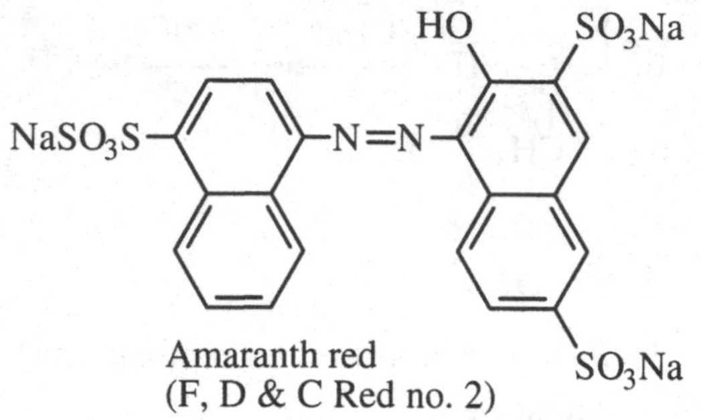

Butter yellow

Amaranth red
(F, D & C Red no. 2)
Banned from food in US since 1976

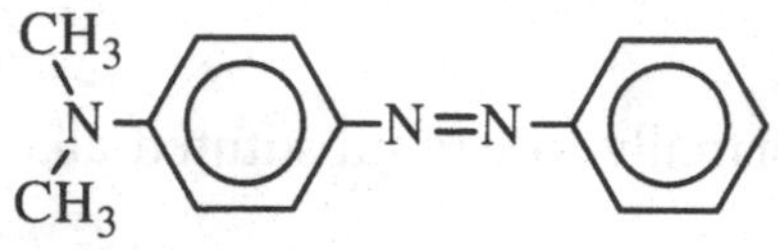

DEMONSTRATION PROBLEM

Propose a synthesis for the pH indicator methyl orange shown in the Application section above. Assume benzene is the only source of aromatic carbons and that you have access to any other needed aliphatic or inorganic reagents.

Solution

This azo dye was clearly prepared by coupling an aryldiazonium with an activated aromatic ring. Because the substituent NaO_3S (or HO_3S if pH is low) is deactivating, that ring must be the one that was part of the aryldiazonium.

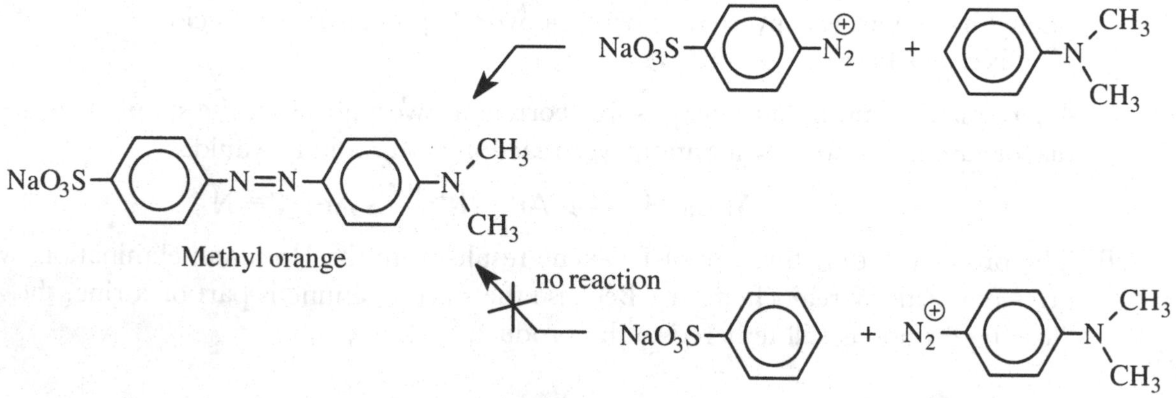

Methyl orange

no reaction

Each component must be made from benzene. The aryldiazonium can come from diazotization of *p*-aminobenzenesulfonic acid that is a sulfonation product of aniline. The other component could also be prepared from aniline by methylation with methyl iodide. Making aniline from benzene is done in two steps by nitration and reduction.

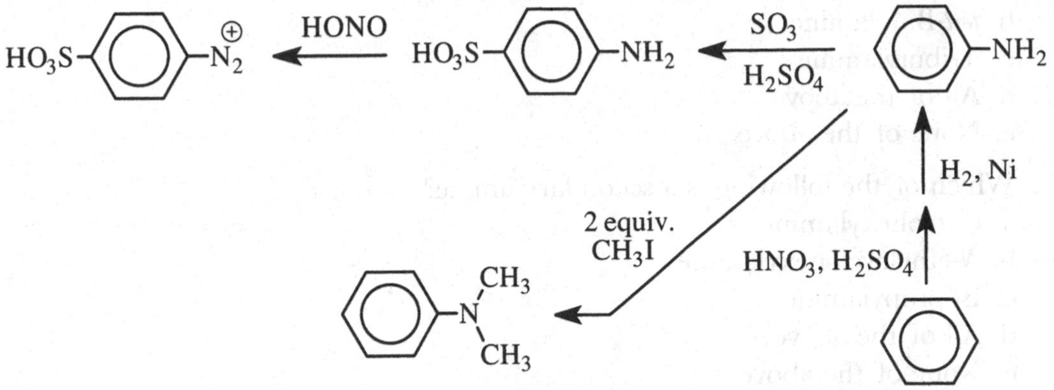

Chapter Test

True/False

1. Morphine contains a tertiary amine (see the structure shown at beginning of the chapter).

2. The pH of an aqueous solution of amine is most likely above 7.

3. If aniline is treated with HONO, H_2SO_4, and then NaI, the final product will be p-iodoaniline.

4. Ph_3N is called tribenzylamine.

5. p-Nitroaniline is a stronger base than cyclohexylamine.

Multiple Choice

6. Which product results from the reaction below?

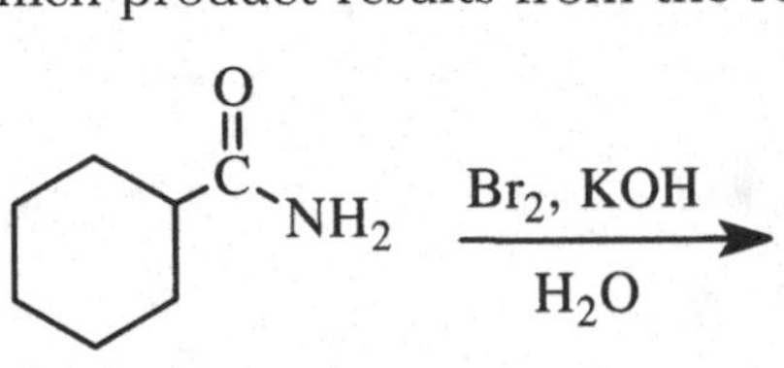

a. 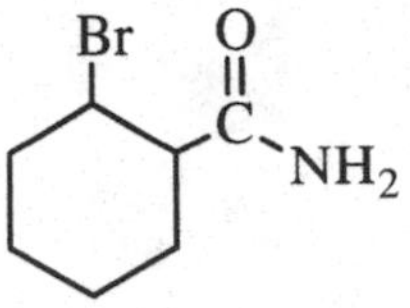   b. 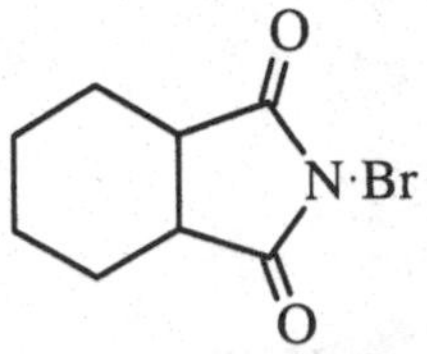   c.

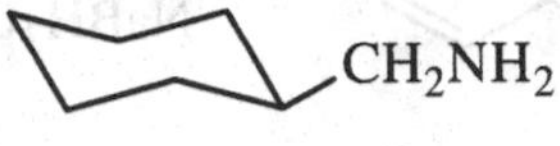

d.    e. None of these

7. 1-Bromooctane could be converted to octylamine by which strategy?
   a. NaN_3, then $LiAlH_4$, then water
   b. Excess CH_3I, then Ag_2O, H_2O, and heat
   c. NaCN, then $LiAlH_4$, then water
   d. Any of the above
   e. None of the above

8. Which of the following can be prepared by a Gabriel amine synthesis?
   a. Aniline

b. *tert*-Butylamine
c. Tributylamine
d. All of the above
e. None of the above

9. Which of the following is a secondary amine?
   a. Cyclohexylamine
   b. *N*-ethylcyclohexylamine
   c. Isopropylamine
   d. All of the above
   e. None of the above

Short Answer

Complete the following with unambiguous structural formulas for the missing organic compounds.

10.

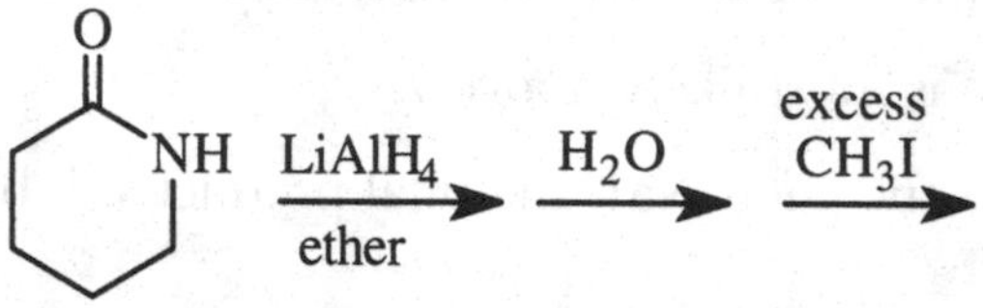

$$\text{(2-piperidinone)} \xrightarrow[\text{ether}]{\text{LiAlH}_4} \xrightarrow{\text{H}_2\text{O}} \xrightarrow[]{\substack{\text{excess}\\ \text{CH}_3\text{I}}}$$

11.

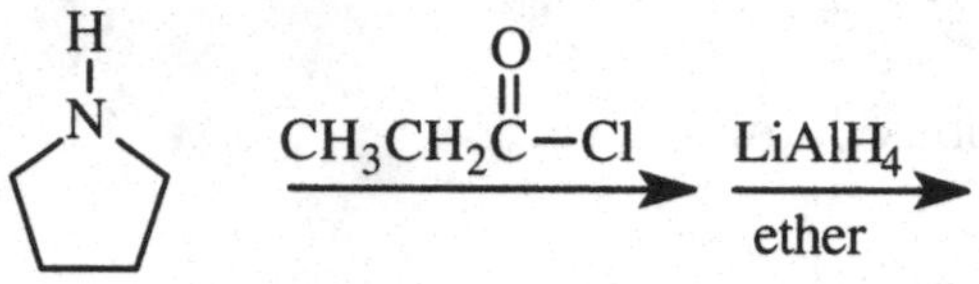

$$\text{(pyrrolidine)} \xrightarrow{\text{CH}_3\text{CH}_2\overset{\text{O}}{\overset{\|}{\text{C}}}-\text{Cl}} \xrightarrow[\text{ether}]{\text{LiAlH}_4}$$

12.

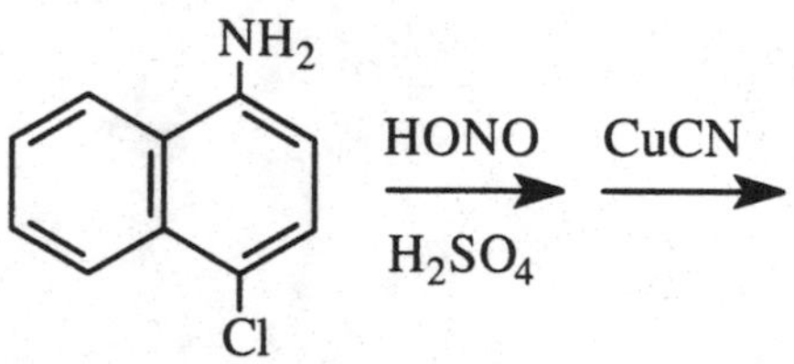

$$\xrightarrow[\text{H}_2\text{SO}_4]{\text{HONO}} \xrightarrow{\text{CuCN}}$$

13.

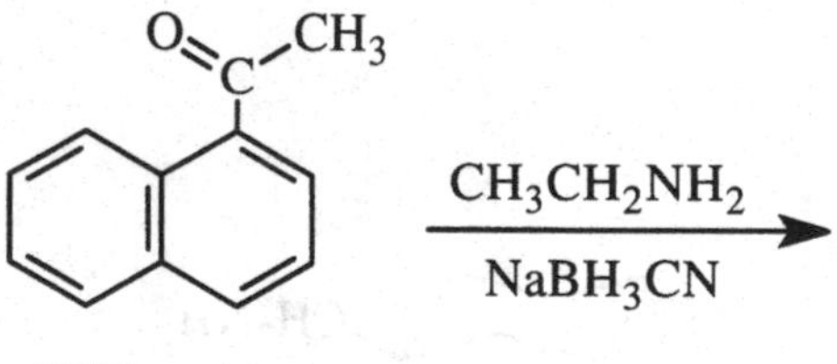

$$\xrightarrow[\text{NaBH}_3\text{CN}]{\text{CH}_3\text{CH}_2\text{NH}_2}$$

14.

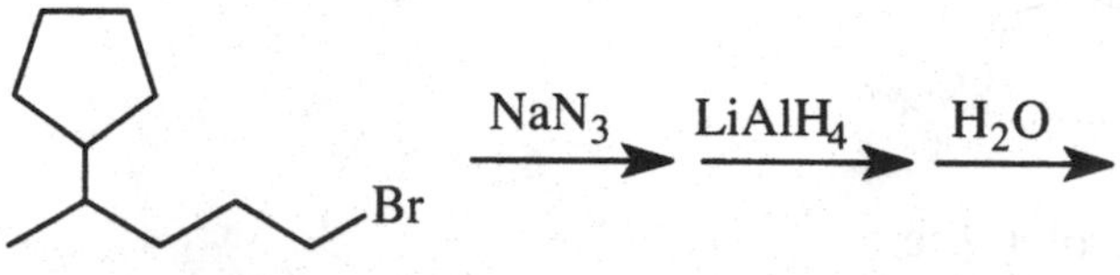

$$\xrightarrow{\text{NaN}_3} \xrightarrow{\text{LiAlH}_4} \xrightarrow{\text{H}_2\text{O}}$$

15.

$$\xrightarrow{\text{NaCN}} \xrightarrow{\text{LiAlH}_4} \xrightarrow{\text{H}_2\text{O}}$$

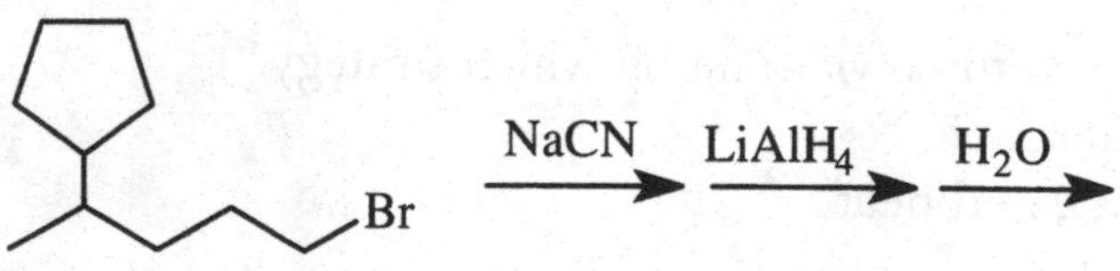

16.

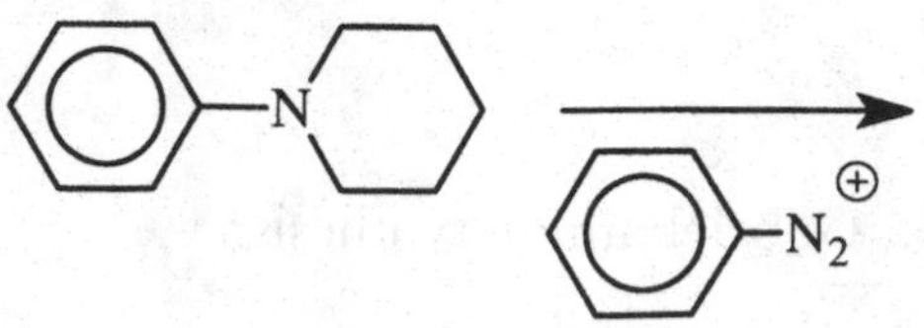

17.

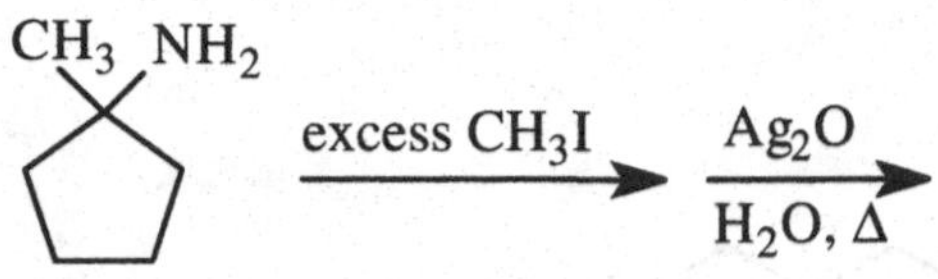

18.

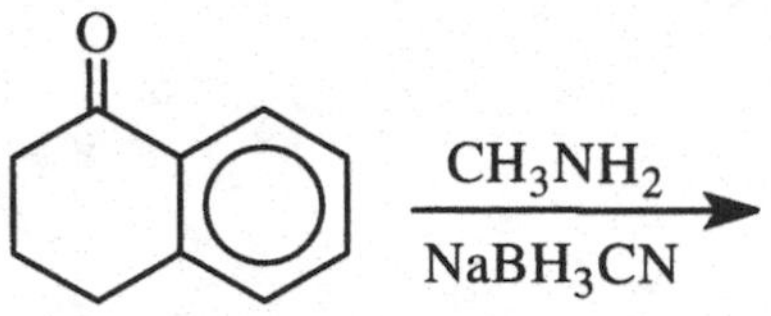

19.

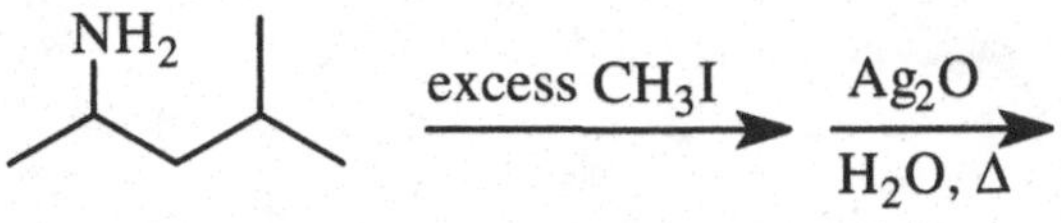

20. What is the product that results when nitrobenzene is treated with Fe, HCl (aq), then HONO, H_2SO_4, and finally H_3PO_2?

21. How one could prepare hexylamine by the Gabriel synthesis.

22. Show the reagents and steps needed to make propylamine from butanoyl chloride by a Curtius rearrangement.

Chapter Test: Answers

1. **True**

2. **True**

3. **False**

4. **False**

5. **False**

6. **d**

7. **a**

8. **e**

9. **b**

10.

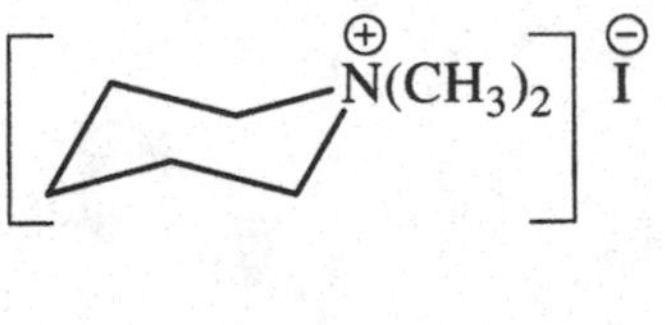

11.

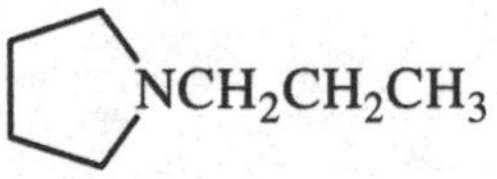

12.

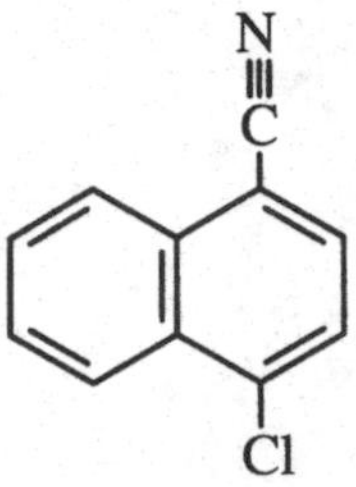

13. 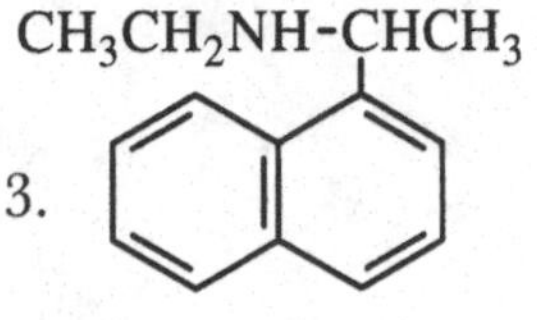 $CH_3CH_2NH-CHCH_3$

14. 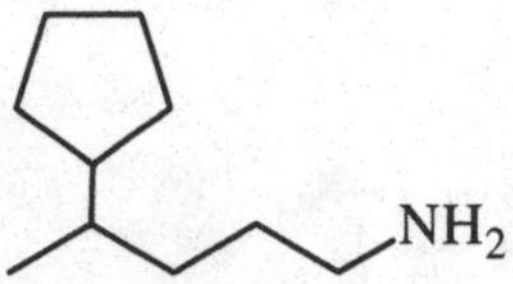 (cyclopentyl with CH-CH₃ chain ending in NH₂)

15. 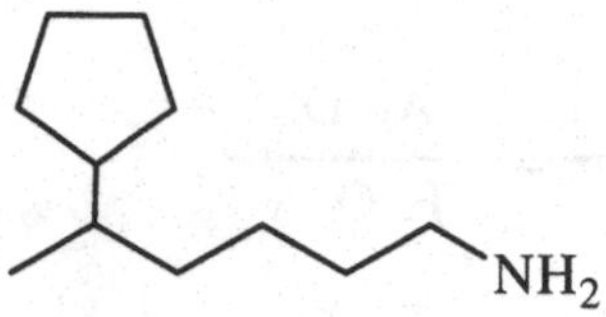 (cyclopentyl chain ending in NH₂)

16. 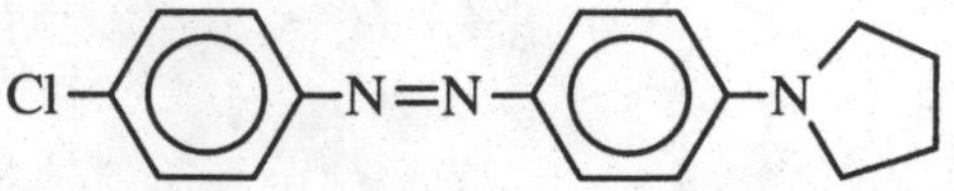 Cl—⟨benzene⟩—N=N—⟨benzene⟩—N⟨pyrrolidine⟩

17. 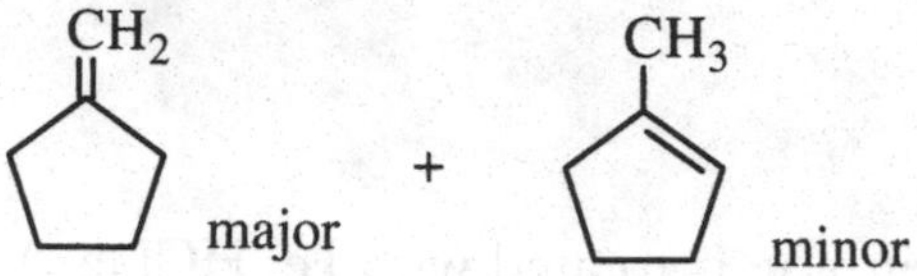 methylenecyclopentane major + 1-methylcyclopentene minor

18. 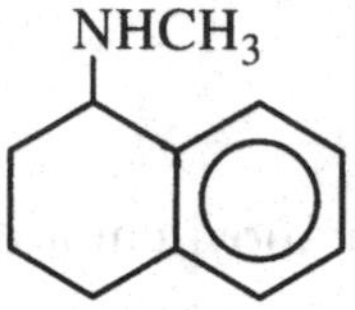 1-(methylamino)tetralin (NHCH₃)

19. 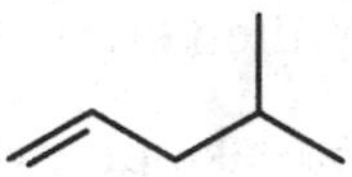 (4-methyl-1-pentene)

20. Benzene

21. Phthalimide $\xrightarrow{\text{KOH}}$ $\xrightarrow{\text{1-bromohexane}}$ $\xrightarrow{\text{NaOH, H}_2\text{O, }\Delta}$

22. $CH_3CH_2CH_2COCl \xrightarrow{\text{NaN}_3} CH_3CH_2CH_2CON_3 \xrightarrow{\Delta}$

$CH_3CH_2CH_2N{=}C{=}O \xrightarrow{\text{H}_2\text{O}} CH_3CH_2CH_2NH_2 + CO_2$

Check Your Performance

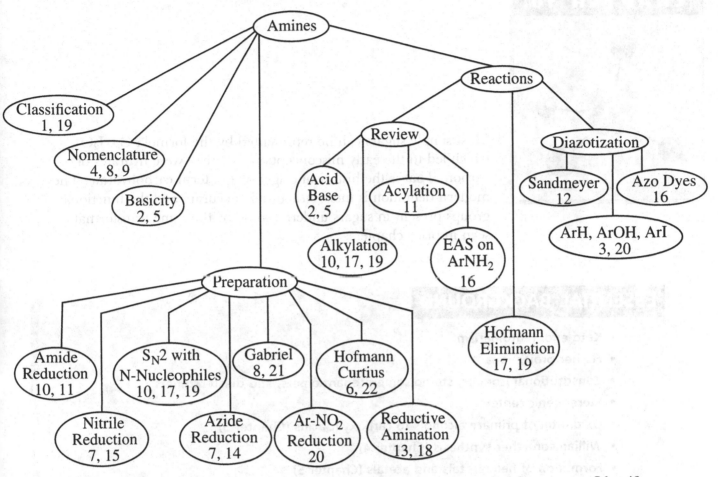

Note the number of questions in each grouping that you got wrong on the chapter test. Identify areas where you need further review and go back to relevant parts of this chapter.

CHAPTER 10

Carbohydrates

Most simple sugars can be represented by the formula $C_n(H_2O)_n$, which led to the early misconception that they were hydrates of carbon. That is the historical origin of the term carbohydrate. The modern definition is more structurally accurate and the functional groups present in sugars undergo some of the same transformations seen in prior chapters.

ESSENTIAL BACKGROUND

- Keto-enol tautomerism
- Fischer projections
- Constitutional isomers, stereoisomers: enantiomers, and diastereomers
- Stereogenic centers
- Oxidation of primary alcohols to carboxylic acids (Chapter 3)
- Williamson ether synthesis (Chapter 4)
- Formation of hemiacetals and acetals (Chapter 5)
- Oxidation of aldehydes (Chapter 5)
- Reduction of aldehydes and ketones (Chapter 5)

TOPIC 1: DEFINITION, STRUCTURE, CLASSIFICATION, AND EXAMPLES

KEY POINTS

✓ *What are carbohydrates?*

✓ *How are carbohydrates classified?*

✓ *What are some important examples of each carbohydrate class?*

A modern definition of **carbohydrate** includes poly-hydroxy aldehydes and ketones or substances that hydrolyze to these. These include **sugars** and many larger molecules made from them. **Simple sugars** are classified as **monosaccharides** because they do not undergo hydrolysis to smaller molecules. **Complex carbohydrates** are those made from two or more monosaccharide units. **Disaccharides, trisaccharides, and polysaccharides** are composed of two, three, or many monosaccharide parts, respectively.

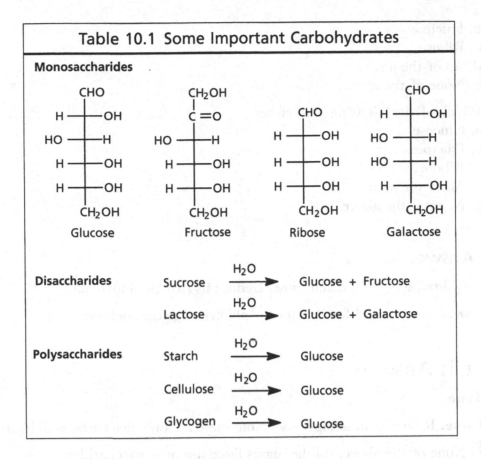

Table 10.1 Some Important Carbohydrates

A common suffix in carbohydrate names and classifications is "ose." A monosaccharide is classified as an **aldose** or a **ketose** depending on whether the functional group present is an aldehyde or a ketone. Classification as a **triose, tetrose, pentose, or hexose** indicates whether the formula contains three, four, five, or six carbon atoms, respectively. These classification methods are often combined such that a given monosaccharide will be an aldotetrose or a ketohexose, etc.

Glucose (blood sugar), fructose (fruit sugar), and ribose (found in RNA) are among the more important monosaccharides. The disaccharide sucrose (table sugar) can be hydrolyzed to glucose and fructose. The disaccharide lactose (milk sugar) contains glucose and galactose. Starch, glycogen, and cellulose are polysaccharides of glucose. These relationships, along with structural formulas for some monosaccharides are shown in **Table 10.1**.

Topic Test 1: Definition, Structure, Classification, and Examples

True/False

1. Carbohydrate names often end in "ose."

2. Ribose (Table 10.1) is an aldohexose.

Multiple Choice

3. Which of the following is a polysaccharide?
   a. Glucose

b. Fructose

c. Ribose

d. All of the above

e. None of the above

4. Which of the following is a ketose?

a. Glucose

b. Fructose

c. Ribose

d. All of the above

e. None of the above

Short Answer

5. Dihydroxyacetone is a ketotriose. Deduce its structural formula.

6. Provide a structural formula for the aldotriose glyceraldehyde.

Topic Test 1: Answers

1. **True**

2. **False.** Ribose is an aldopentose. Note there are only five carbons in its structure.

3. **e.** None of the above. All the sugars listed are monosaccharides.

4. **b.** Fructose. Glucose and ribose contain the aldehyde functional group and are therefore classified as aldose sugars.

5. A ketotriose is a three-carbon ketone with the formula $C_3(H_2O)_3$. The only reasonable stable structure is, as the name dihydroxyacetone implies, a three-carbon chain with a carbonyl in the middle and hydroxy groups on the ends.

$$\underset{}{\overset{\displaystyle O}{\underset{\displaystyle \|}{}}}$$
$$HOCH_2CCH_2OH$$

6. An aldotriose is a three-carbon aldehyde with the formula $C_3(H_2O)_3$. The correct answer has the IUPAC name 2,3-dihydroxypropanol.

$$\underset{\displaystyle \underset{OH\quad OH}{|\quad\ |}}{CH_2-CH-\overset{\displaystyle \overset{O}{\|}}{C}-H}$$

TOPIC 2: STEREOCHEMISTRY OF MONOSACCHARIDES

KEY POINTS

✓ *What is the conventional orientation of a Fischer projection for a monosaccharide?*

✓ *How is the configuration of a sugar defined as D or L?*

✓ *What is the configuration of naturally occurring sugars?*

✓ *How does one identify relationships between stereoisomeric sugars?*

Fischer projections are used extensively to represent monosaccharides. The conventional orientation for a Fischer projection of a monosaccharide puts carbons in a vertical line with the carbonyl (aldehyde or ketone) as close to the top as possible. The monosaccharides shown in Table 10.1 follow this convention.

Enantiomeric forms of a given sugar will have the same name except for a prefix "D" or "L" to indicate which enantiomer it is. When a Fischer projection in its conventional orientation has the OH of the bottom stereogenic center on the right, the sugar is designated "D." If that OH is on the left, the designator "L" is used. These are based on the structural analogy to the simplest chiral sugar, glyceraldehyde.

D-Glyceraldehyde   L-Glyceraldehyde

The designators D and L do not, in general, indicate anything about optical rotation in any case except for glyceraldehyde itself. Although D-glyceraldehyde is dextrorotatory, we cannot conclude that all D sugars will be. Essentially all naturally occurring monosaccharides have the D configuration. Note also that enantiomers have opposite configurations around all stereogenic centers and not just the bottom one. If only one or some of the chiral centers are opposite, the stereoisomers are diastereomers as illustrated by four aldotetrose isomers.

D-Erythrose   L-Erythrose   D-Threose   L-Threose

Topic Test 2: Stereochemistry of Monosaccharides

True/False

1. Monosaccharides found in nature usually have the D configuration.

2. L sugars are designated as such because they are levorotatory.

Multiple Choice

3. Consult the structures of glucose and galactose shown in Table 10.1. The structures shown
   a. both have the D configuration.
   b. both have the L configuration.
   c. are one of each D and L.
   d. are both achiral.
   e. None of the above

4. The relationship between the galactose and glucose shown in Table 10.1 is that they are
   a. enantiomers.
   b. diastereomers.
   c. constitutional isomers.
   d. identical.
   e. None of the above

Short Answer

5. The structure of D-ribose is shown in Table 10.1. Use it to deduce the structure of L-ribose.

Topic Test 2: Answers

1. **True**

2. **False.** The L designation is used to express the similarity of structure to the chiral aldose L-glyceraldehyde (which is levorotatory). Other L sugars might be dextrorotatory or levorotatory, and there is no easy way to predict which will be observed.

3. **a.** Both have the D configuration.

4. **b.** Diastereomers. Note that the structures are the same in every respect except the configuration of C2. These structures have the same formula and order of linkage so they must be stereoisomers. They are not mirror images and cannot, therefore, be enantiomers so they must be diastereomers.

5. L-ribose will be the mirror image of D-ribose (i.e., it will have the opposite absolute configuration around all stereogenic centers). (Do not make the common mistake of drawing a structure in which only the C4 stereochemistry is opposite. That does give an L sugar but it is not L-ribose.)

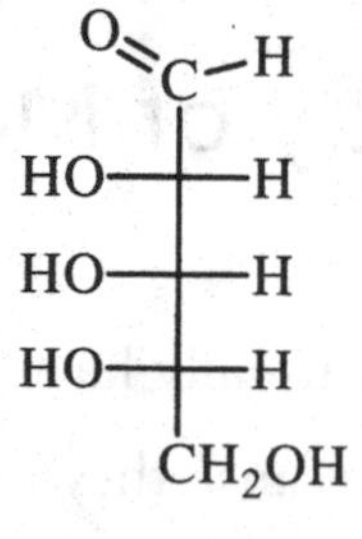

L-Ribose

TOPIC 3: CYCLIC HEMIACETALS AND HAWORTH PROJECTIONS

KEY POINTS

✓ *How do monosaccharides form cyclic hemiacetals?*

✓ *What are Haworth projections and what conventions are used for drawing them?*

✓ *What are anomers and how are they named and classified?*

✓ *What are the consequences of the equilibrium between open chain and cyclic forms?*

Recall from Chapter 5 that alcohols add reversibly to the carbonyls of aldehydes or ketones to yield hemiacetals. The reaction can be intramolecular where the reacting alcohol and carbonyl are within the same molecule and the hemiacetal product will be cyclic.

$$R-\overset{\overset{\displaystyle O}{\|}}{C}-H\,(R) \;+\; HOR' \;\;\rightleftharpoons\;\; R-\overset{\overset{\displaystyle OH}{|}}{\underset{\underset{\displaystyle OR'}{|}}{C}}-H\,(R)$$

Hemiacetal

Many monosaccharides form cylic hemiacetals. The ring can be five or six membered and is called a **furanose** or **pyranose**, respectively. Although the rings are actually puckered, they are often drawn as planar in pictures, called **Haworth projections**, in which the ring is viewed from slightly above a horizontal perspective. The ring oxygen is placed at the 12 o'clock or 2 o'clock position and the former carbonyl carbon (called the **anomeric carbon**) is positioned at the right if possible.

A Pyranose Skeleton

A Furanose Skeleton

The anomeric carbon is stereogenic as a result of the ring closure. There are two possible configurations leading to two different stereoisomers, more specifically called **anomers**. Anomers are given the prefix designators α or β depending on the orientation of the hydroxyl group on the anomeric carbon. For D sugars, a conventional Haworth projection of the α anomer shows the anomeric hydroxyl pointed down (i.e., *trans* to the C6 hydroxymethyl, CH_2OH, in the case of glucose). In aqueous solution there is a dynamic equilibrium between the open chain form and the two anomers of a monosaccharide. A pure sample of any of the three will equilibrate to a mixture if placed in solution. These points are illustrated with the glucose example below.

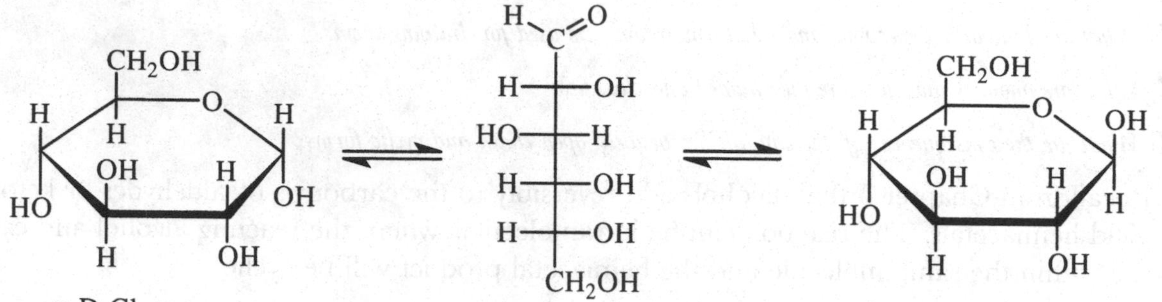

α-D-Glucopyranose   D-Glucose   β-D-Glucopyranose

Topic Test 3: Cyclic Hemiacetals and Haworth Projections

True/False

1. Placing a sample of the pure α-D-glucopyranose in water will make it isomerize to a sample of pure β-D-glucopyranose.

2. An aldotetrose cannot form a pyranose ring.

Multiple Choice

3. The α and β forms of glucopyranose are
   a. stereoisomers.
   b. diastereomers.
   c. anomers.
   d. All of the above
   e. None of the above

4. Which of the following are anomers?
   a. β-D-glucopyranose and β-L-glucopyranose
   b. β-D-glucopyranose and β-D-glucofuranose
   c. β-D-glucopyranose and α-D-glucopyranose
   d. All of the above
   e. None of the above

Short Answer

5. Use the structure of D-fructose shown in Table 10.1 to deduce the structure of β-D-fructofuranose.

6. A conventional Haworth projection of a particular monosaccharide is shown below. Is the open chain form of the monosaccharide an aldose or a ketose? How can one tell?

Topic Test 3: Answers

1. **False.** An isomerization will occur but it will be a dynamic equilibrium containing both anomers along with a small amount of the open chain form.

2. **True.** To form a six-memebered ring hemiacetal, the sugar must have an oxygen located on the fifth carbon away from the carbonyl. An aldotetrose does not have an alcohol oxygen appropriately positioned to make a pyranose. (Try it and see.)

3. **d.** All of the above. Although the discussion in this topic clearly states that the α and β forms are anomers (i.e, isomers differing around the anomeric carbon), it is also true that these anomers are diastereomers that are a subset of stereoisomers.

4. **c.** β-D-glucopyranose and α-D-glucopyranose. These are the two that differ by stereochemistry around C1. The others differ by configuration around all stereogenic centers or the size of the ring.

5.

β-D-Fructofuranose

6. The monosaccharide shown is an aldose. Note that the anomeric carbon (which was the carbonyl in the open chain form) bears a hydrogen atom. That means it must have been an aldose because a ketose would have two carbons bound to the anomeric carbon.

TOPIC 4: SOME REACTIONS OF MONOSACCHARIDES

KEY POINTS

✓ *How are the OH groups of carbohydrates acylated?*

✓ *How are the OH groups of a carbohydrate methylated?*

✓ *What are glycosides and how are they classified and named?*

Many reactions of monosaccharides resemble analogous transformations discussed for alcohols (Chapter 3) or aldehydes and ketones (Chapter 5). The hydroxyl groups can act as nucleophiles for an acyl substitution on the carbonyl of an anhydride or acyl chloride. Acetic anhydride is often used to convert the alcohol oxygens to acetate esters.

If the monosaccharide is treated with methyl iodide in the presence of silver ion, a silver-assisted reaction resembling the Williamson ether synthesis will methylate all the hydroxyl groups.

Recall from Chapter 5 that hemiacetals react with alcohols to make acetals. The cyclic hemiacetals of monosaccharides behave similarly to yield acetals more commonly called **glycosides**. Once the OH of the anomeric carbon is replaced by an OR to make a glycoside, there is no longer an equilibrium between the anomers and the open chain form. The glycoside can be hydrolyzed back to the hemiacetal with aqueous acid.

Like anomers, glycosides can be α or β depending on the direction of the OR group. Although the alcohol used to make the glycoside from the hemiacetal can be simple like methanol or ethanol, in nature the alcohol is usually another sugar molecule. Glycoside linkages connect monosaccharide units to form di-, tri-, and polysaccharides. The glycosides are designated by numbers and Greek letters to indicate which carbon of one sugar is linked via the glycoside to which carbon on the other sugar. One number is "primed" (indicated as n′) to specify there are two different sugars. For example, cellobiose and maltose are each disaccharides of glucose that are held together by α-1,4′ and β-1,4′ glycosides, respectively.

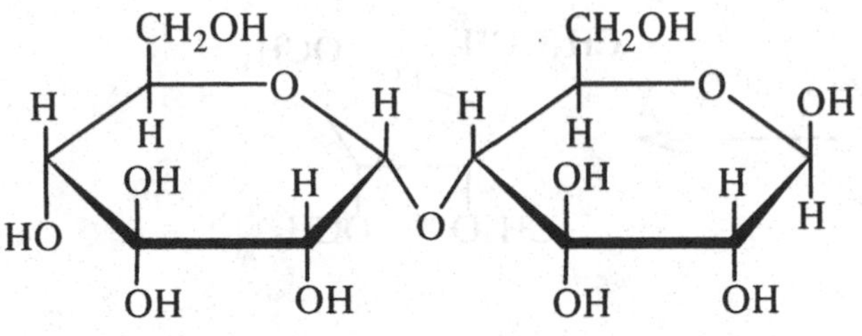

Maltose
an α-1,4′ glycoside

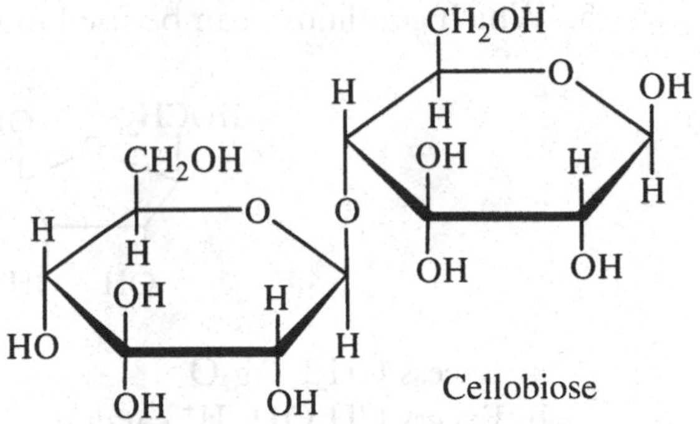

Cellobiose
a β-1,4′ glycoside

Topic Test 4: Some Reactions of Monosaccharides

True/False

1. Glycosides are actually hemiacetals.

2. A disaccharide is made of two monosaccharides connected by a glycoside linkage.

Multiple Choice

3. Which conditions can be used to carry out the transformation below?

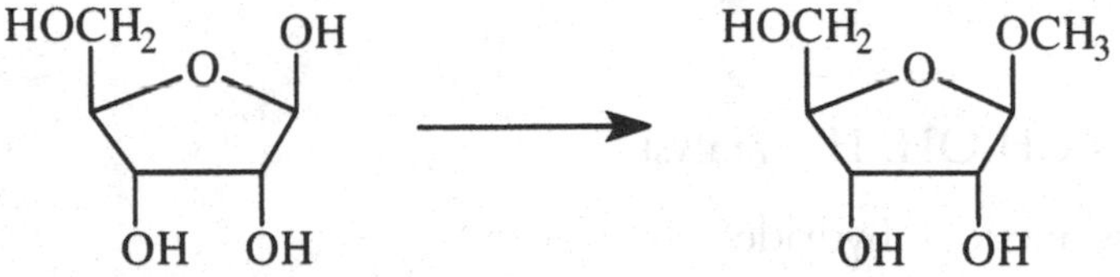

   a. Excess CH_3I, Ag_2O
   b. Excess CH_3OH, H^+ catalyst
   c. Excess acetic anhydride
   d. Excess dry acid catalyst
   e. None of the above

4. Which conditions can be used to carry out the transformation below?

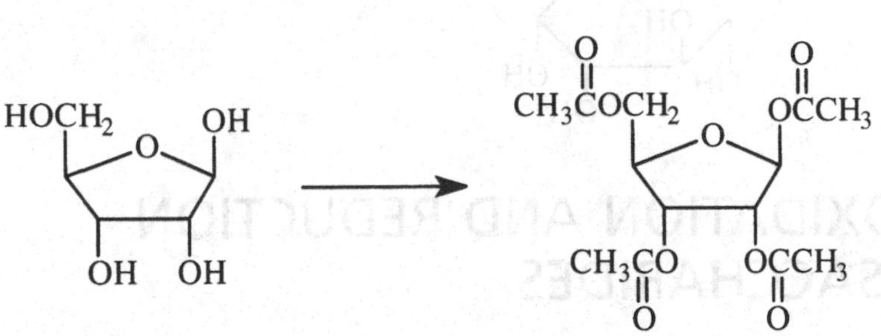

   a. Excess CH_3I, Ag_2O
   b. Excess CH_3OH, H^+ catalyst
   c. Excess acetic anhydride
   d. Excess dry acid catalyst
   e. None of the above

5. Which conditions can be used to carry out the transformation below?

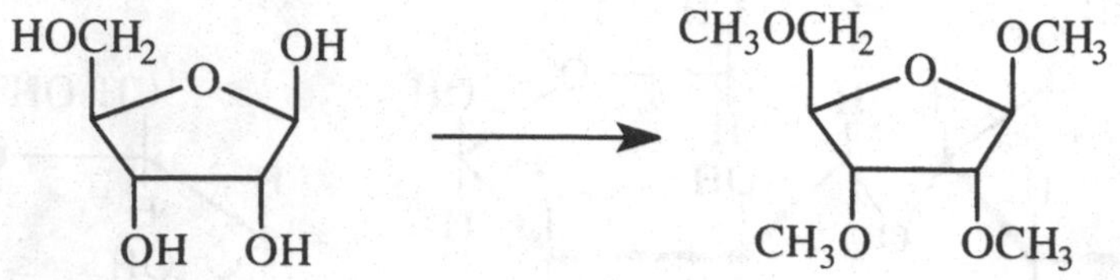

   a. Excess CH_3I, Ag_2O
   b. Excess CH_3OH, H^+ catalyst
   c. Excess acetic anhydride
   d. Excess dry acid catalyst
   e. None of the above

Short Answer

6. Draw a β-1,6' gycoside of glucose.

Topic Test 4: Answers

1. **False.** Glycosides are actually acetals that formally result from cyclic hemiacetal sugars reacting with alcohols.

2. **True**

3. **b.** Excess CH_3OH, H^+ catalyst.

4. **c.** Excess acetic anhydride

5. **a.** Excess CH_3I, Ag_2O

6.

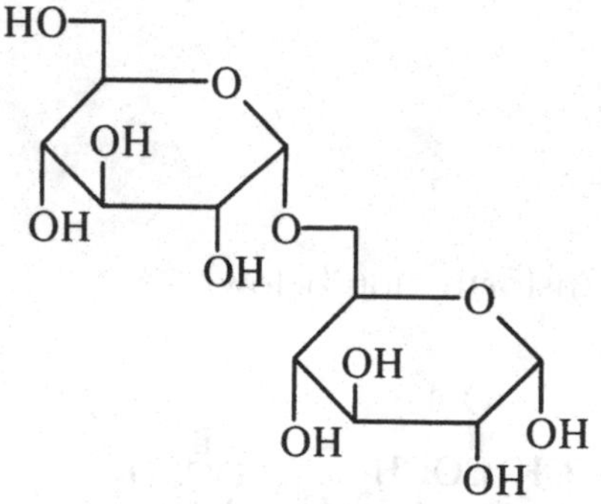

TOPIC 5: OXIDATION AND REDUCTION OF MONOSACCHARIDES

KEY POINTS

✓ *What is a reducing sugar?*

✓ *What products result from reaction with Tollens', Benedict's, or Fehling's reagent?*

✓ *What products result from reaction with bromine and water?*

✓ *What products result from reaction with nitric acid?*

✓ *What products result from reaction with sodium borohydride?*

Recall from Chapter 5 that **Tollens'** reagent selectively oxidizes aldehydes to carboxylates (or carboxylic acids when the pH is lowered). Similar results can be obtained from **Benedict's** or **Fehling's** reagents that contain Cu^{2+} as the oxidizing agent. Aqueous bromine (Br_2, H_2O) can also carry out this selective oxidation of aldehydes.

$$R-\overset{\overset{O}{\parallel}}{C}-H \xrightarrow[\substack{\text{or } Cu^{2\oplus} \\ \text{Fehling's or Benedict's} \\ \text{reagent}}]{\substack{Ag(NH_3)_2^{\oplus} \\ \text{(Tollens' reagent)}}} RCO_2^{\ominus} \xrightarrow{H_3O^{\oplus}} RCO_2H$$

$$Br_2, H_2O$$

Sugars that react in this way are called **reducing sugars** because they reduce the metal ion. An aldose, or anything that provides a free aldehyde group in equilibrium, will be a reducing sugar. Cyclic hemiacetals and some ketoses that equilibrate with aldehydes via keto-enol tautomerism are reducing sugars even though the concentration of aldehyde is low. In these cases the small amount of aldehyde reacts and then is replaced by more that also reacts, and so on (Le Châtelier's principle) until all the sugar is oxidized. Many glycosides are not reducing sugars because the aldehyde is not present in equilibrium.

More vigorous oxidation with warm aqueous nitric acid will convert both an aldehyde and primary alcohol of a monosaccharide to carboxylic acids while leaving the secondary alcohols unchanged.

The carbonyl of a monosaccharide can be reduced to an alcohol with sodium borohydride.

Topic Test 5: Oxidation and Reduction of Monosaccharides

True/False

1. The aldehyde of an aldose can be reduced to a primary alcohol with warm HNO_3.

2. Ketoses do not contain an aldehyde and therefore are not reducing sugars.

Multiple Choice

3. Which conditions below will oxidize an aldose to a carboxylic acid?
   a. $Ag(NH_3)_2^+$, then H_3O^+
   b. Cu^{2+}, then H_3O^+
   c. HNO_3, warm
   d. All of the above
   e. None of the above

4. Which of the following is *not* a reducing sugar?

   d. All of the above
   e. None of the above

Short Answer

Provide a structure for the missing organic compound.

5.

H—C=O
H——OH
H——OH
CH₂OH

$\xrightarrow{\text{NaBH}_4}$ $\xrightarrow{\text{H}_2\text{O}}$

6. The structure of sucrose is shown in the Application section following this topic test. Is sucrose a reducing sugar? Explain.

Topic Test 5: Answers

1. **False.** Treating an aldose with warm nitric acid will oxidize both the aldehyde and the primary alcohol on the last carbon of the chain into carboxylic acids.

2. **False.** Many ketoses are in equilibrium with an aldehyde via keto-enol tautomesism, which allows them to react with Tollens' reagent. The ketose fructose, for example, is a reducing sugar.

3. **d.** All of the above. The first two are Tollens' and Benedict's or Fehling's reagents that act only on the aldehyde. The warm nitric acid will oxidize both the aldehyde and the primary alcohol on the last carbon of the chain.

4. **c.** The glycoside in answer c is not a reducing sugar because it is not in equilibrium with an open chain aldehyde and therefore cannot react with Tollens', Benedict's, or Fehling's reagents. Answers a and b are an open chain aldose and the hemiacetal of an aldose that equilibrates with an open chain aldose.

5.

CH₂OH
H——OH
H——OH
CH₂OH

6. Sucrose is not a reducing sugar because both anomeric carbons are "tied up" in the glycoside and therefore there is no equilibration with an open chain aldehyde.

APPLICATION

Sucrose (table sugar) is a disaccharide of glucose and fructose. The average annual sucrose consumption in the United States is about 150 pounds per person. Although pleasant to the taste, sucrose does have some undesirable dietary properties. Sucrose metabolism involves hydrolysis to glucose and fructose.

Metabolism of the glucose is regulated by the protein hormone insulin that controls the level of glucose in the blood. Diabetics do not produce sufficient insulin and must restrict their glucose and sucrose intake to avoid increased blood glucose levels. Sucrose is converted by bacteria found in the mouth into a material called dextram that makes up about 10% of the dental plaque that leads to tooth decay. Reduction of the aldehyde of glucose yields glucitol (also called sorbitol).

Sorbitol is a sweetener used in some hard candies and gums. Although sorbitol is not low calorie, it is a desirable sucrose substitute in some applications because its metabolism is not insulin regulated and it also cannot be converted by bacteria into tooth-decaying dextram.

DEMONSTRATION PROBLEM

Allose is an aldohexose with a structure similar to glucose except for an opposite stereochemical configuration at C3. Draw a Fischer projection of D-allose and then use that to deduce a Haworth projection of the β anomer of D-allose in its pyranose form. Draw a disaccharide of D-allose that contains a β-1,3′ glycosidic link.

Solution

From the description of allose given in the problem, we can deduce the Fischer projection of D-allose based on the structrue of D-glucose. Note that the aldohexose is the same everywhere except at carbon 3, which has the opposite stereochemical configuration. The Fischer formula of D-allose is then converted to a Haworth form by closing a six-membered ring cyclic hemiacetal between the carbonyl and the OH on C5 as shown below. The OH on C1 will be *cis* to C7 (pointing up) because the β anomer was specified.

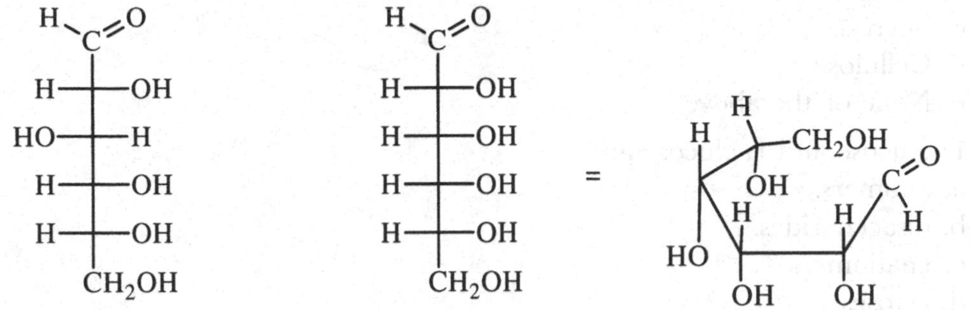

D-Glucose   D-Allose

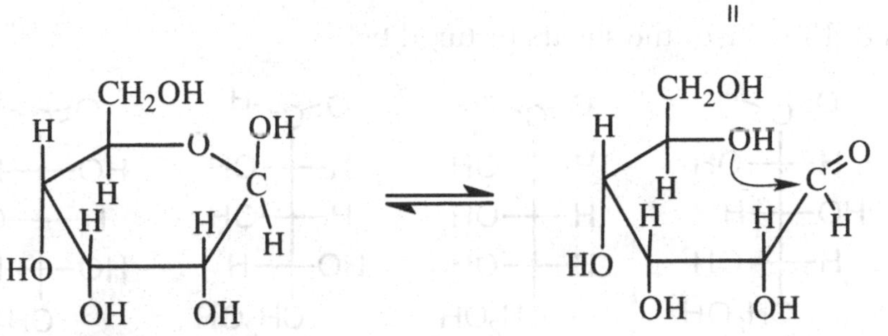

β-D-Allopyranose

The β-1,3′ glycoside will have two allopyranose rings connected via an oxygen between C1 (the anomeric carbon) of one allose and C3′ of the other.

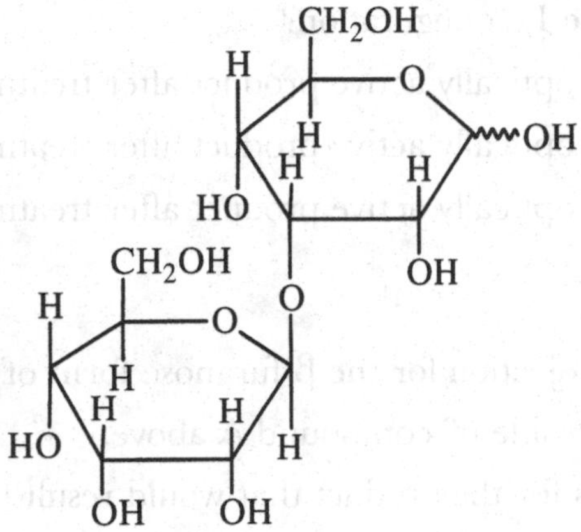

Chapter Test

True/False

1. Anomers are stereoisomers.

2. Aldoses are reducing sugars.

3. A six-membered ring is called a furanose.

4. Tollens' reagent contains Cu^{2+}.

5. Polysaccharides are monosaccharides held together by hydrogen bonding.

Multiple Choice

6. Which of the following is a polysaccharide?
   a. Glucose
   b. Ribose
   c. Sucrose
   d. Cellulose
   e. None of the above

7. D-glucose and L-glucose are
   a. anomers.
   b. disaccharides.
   c. enatiomers.
   d. ketoses.
   e. None of the above

Problems 8–13 refer to the sugars pictured below.

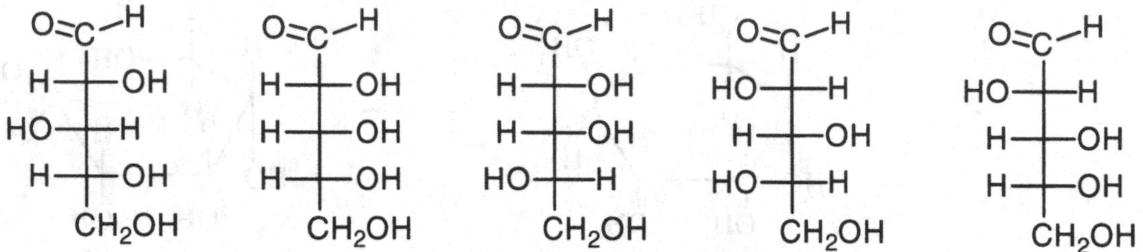

8. Which is (are) optically active?

9. Which are enantiomers of one another?

10. Which has (have) the L configuration?

11. Which will yield an optically active product after treatment with $NaBH_4$ and water?

12. Which will yield an optically active product after treatment with bromine and water?

13. Which will yield an optically active product after treatment with warm nitric acid?

Short Answer

14. Draw a Haworth projection for the β furanose form of compound **A** above.

15. Draw an α-1,5′ glycoside of compound **A** above.

Provide structural formulas for the product that would result from treating the compound below with each of the following reagents.

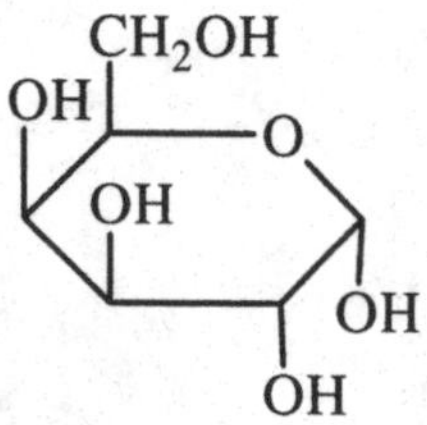

16. Excess acetic anhydride

17. Excess CH_3I, Ag_2O

18. Excess CH_3CH_2OH with dry acid catalyst

19. Convert the Haworth projection shown as the reactant for problems 16–18 into a Fischer projection of the open chain monosaccharide.

20. Classify the reactant shown above for problems 16–18 as an α or β anomer; a furanose or a pyranose; a D or L sugar; an aldose or ketose; and a triose, tetrose, pentose, or hexose.

Chapter Test: Answers

1. **True**
2. **True**
3. **False**
4. **False**
5. **False**
6. **d**   7. **c**
8. **All**
9. **A** and **D**
10. **C** and **D**
11. **C** and **E**
12. **A, B, C, D**, and **E**
13. **C** and **E**

14. 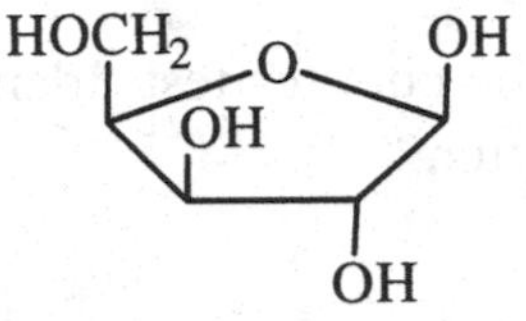 [Haworth structure: HOCH₂, O, OH, OH, OH]

15. 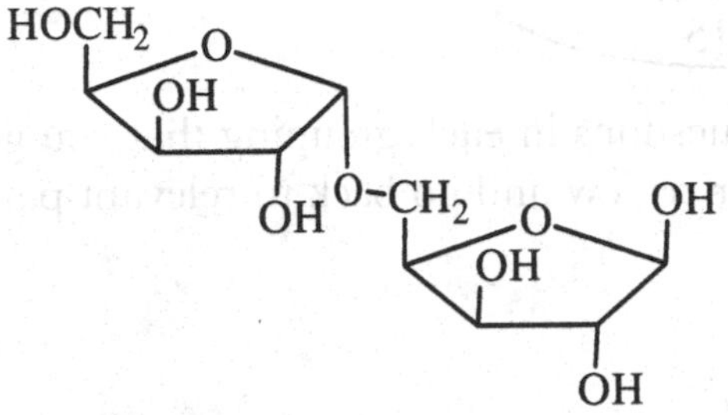 [Haworth structure: HOCH₂, O, OH, OH, O–CH₂, O, OH, OH, OH]

16. 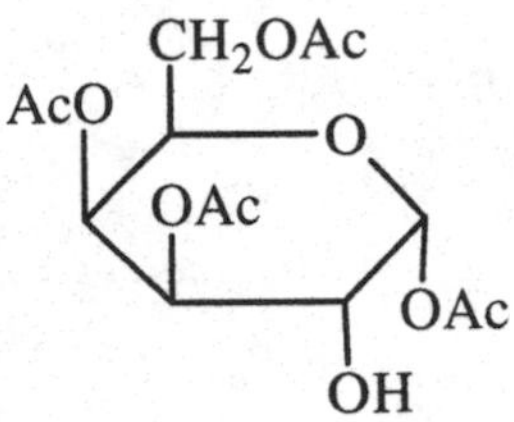 [Haworth structure: AcO, CH₂OAc, O, OAc, OAc, OH]
(Ac = acetyl = CH₃C(=O)—)

17. 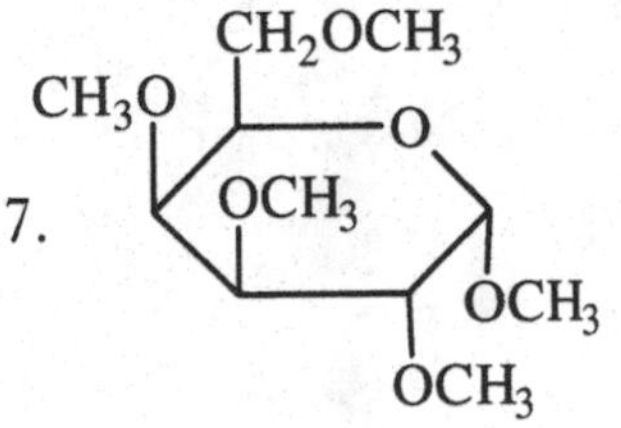 [Haworth structure: CH₃O, CH₂OCH₃, O, OCH₃, OCH₃, OCH₃]

18. 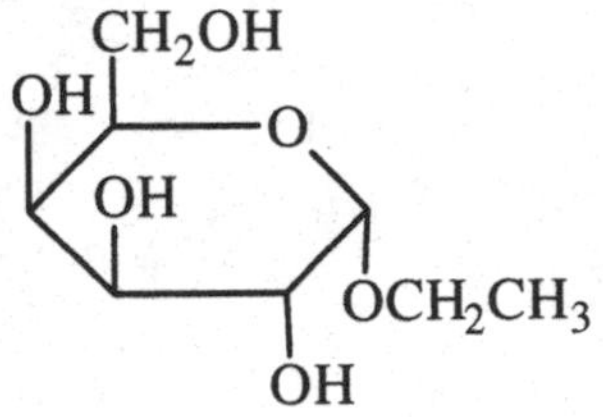 [Haworth structure: OH, CH₂OH, O, OH, OCH₂CH₃, OH]

19. 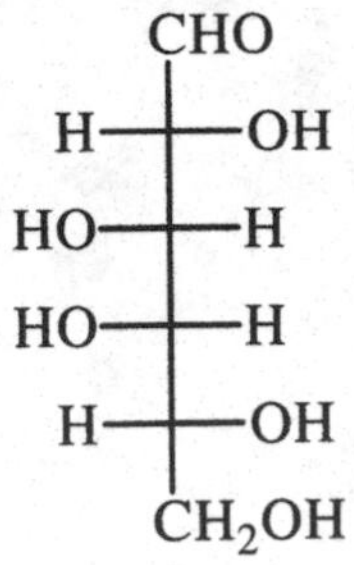
$$
\begin{array}{c}
\text{CHO} \\
\text{H}\!-\!\!-\!\text{OH} \\
\text{HO}\!-\!\!-\!\text{H} \\
\text{HO}\!-\!\!-\!\text{H} \\
\text{H}\!-\!\!-\!\text{OH} \\
\text{CH}_2\text{OH}
\end{array}
$$

20. It is the α-pyranose form of a D-aldohexose.

Check Your Performance

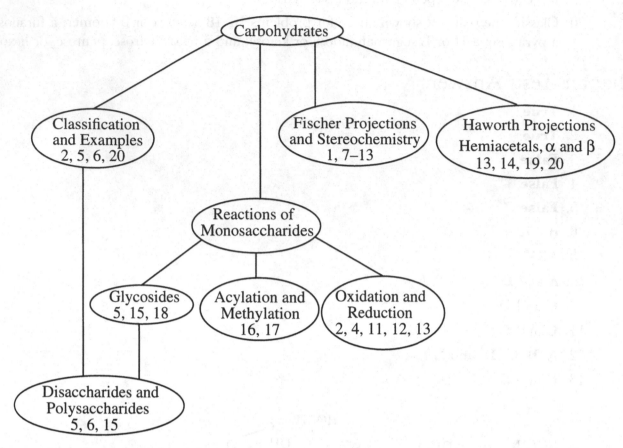

Note the number of questions in each grouping that you got wrong on the chapter test. Identify areas where you need further review and go back to relevant parts of this chapter.

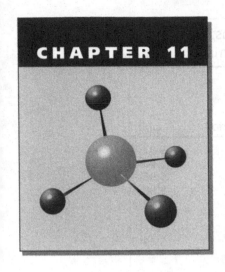

CHAPTER 11

Amino Acids, Peptides, and Proteins

Proteins make up about half of the dry weight (i.e., excluding water) and about 15% of the wet weight of the human body. They are found in greater variety and perform a greater number of biological functions than any other class of compounds. They include enzymes that catalyze most biochemical reactions, transport proteins such as hemoglobin, and the structural proteins that make up skin, hair, and fingernails. In this chapter we begin by examining amino acids and then explore how these are connected to form peptides and proteins.

ESSENTIAL BACKGROUND

- Oxidation of thiols to disulfides (Chapter 3)
- Acid-base reactions of carboxylic acids (Chapter 6)
- Preparation and hydrolysis of amides (Chapter 7)
- Nucleophilic acyl substitution (Chapter 7)
- Fischer esterification and ester hydrolysis (Chapter 7)
- Nitrile hydrolysis (Chapter 7)
- HVZ reaction: alpha-bromination of carboxylic acids (Chapter 8)
- Acid-base reactions of amines (Chapter 9)
- Gabriel synthesis of amines (Chapter 9)

TOPIC 1: STRUCTURE, CLASSIFICATION, NOMENCLATURE, AND SYNTHESIS OF AMINO ACIDS

KEY POINTS

✓ *What is the general structure of an amino acid?*

✓ *What are the names and structures of the common naturally occurring amino acids?*

✓ *What are the three-letter abbreviations used for amino acids?*

✓ *How are amino acids classified according to structure?*

✓ *How are amino acids synthesized in the laboratory?*

Amino acids are both amines and acids. Specifically, they are alpha amino carboxylic acids and are often represented by the general formula shown in **Table 11.1**. Nearly all naturally

Table 11.1 Structures, Names, Abbreviations, and Isoelectric Points of Common Amino Acids

$$H_2N-CH-CO_2H$$
$$|$$
$$R$$

SIDE CHAIN, R	NAME	ABREVIATION	pI
$-H$	Glycine	Gly	5.97
$-CH_3$	Alanine	Ala	6.00
$-CH(CH_3)_2$	Valine	Val	5.96
$-CH_2CH(CH_3)_2$	Leucine	Leu	5.98
$-CHCH_2CH_3$ $\quad|$ $\quad CH_3$	Isoleucine	Ile	6.02
$-CH_2Ph$	Phenylalanine	Phe	5.48
$-CH_2$ (indole ring structure)	Tryptophan	Trp	5.89
$-CH_2OH$	Serine	Ser	5.68
$-CHOH$ $\quad|$ $\quad CH_3$	Threonine	Thr	5.60
$-CH_2$—(phenol ring)—OH	Tyrosine	Tyr	5.66
$-CH_2CO_2H$	Aspartic acid	Asp	2.77
$-CH_2CH_2CO_2H$	Glutamic acid	Glu	3.22
$-CH_2CONH_2$	Asparagine	Asn	5.41
$-CH_2CH_2CONH_2$	Glutamine	Gln	5.65
$-CH_2CH_2CH_2CH_2NH_2$	Lysine	Lys	9.74
$-CH_2CH_2CH_2NHCNH_2$ (with NH double bond)	Arginine	Arg	10.76
$-CH_2$ (imidazole ring structure)	Histidine	His	7.59
$-CH_2SH$	Cysteine	Cys	5.07
$-CH_2CH_2SCH_3$	Methionine	Met	5.74
(pyrrolidine ring structure with $CO_2^\ominus$ and $NH_2^\oplus$) (complete structure)	Proline	Pro	6.30

occurring amino acids have this form and differ only by the identity of the side chain R. All except proline are primary amines. All except glycine are stereogenic at the alpha carbon. The names, side chains, and common three-letter abbreviation for 20 naturally occurring amino acids are shown in Table 11.1.

Depending on the nature of its side chain, a given amino acid might be classified as acidic, basic, **hydrophilic** (water loving), **hydrophobic** (water fearing), aromatic, or sulfur-containing. A specific amino acid might belong to two or more of these categories.

Laboratory synthesis of a given amino acid can be from the corresponding carboxylic acid, which is alpha brominated (HVZ reaction, Chapter 8) and then reacted with excess ammonia or phthalimide ion in a Gabriel synthesis (Chapter 9).

$$RCH_2CO_2H \xrightarrow[\text{2) } H_2O]{\text{1) } Br_2, PBr_3} \underset{\underset{Br}{|}}{RCHCO_2H} \xrightarrow{\overset{\text{excess}}{NH_3}} \underset{\underset{NH_2}{|}}{RCHCO_2H}$$

Amino acids are also available from aldehydes having one fewer carbon via a process called the **Strecker synthesis**. The aldehyde is first converted to an amino nitrile that is then hydrolyzed. Note that the former aldehyde carbon becomes the alpha carbon in the product and that the cyanide ion provides what ultimately becomes the carboxyl carbon.

$$\underset{R}{\overset{\overset{\displaystyle O}{\|}}{C}}\diagup^{H} \xrightarrow[H_2O]{NH_4Cl, KCN} \underset{\underset{NH_2}{|}}{R-CH-C\equiv N} \xrightarrow{H_3O^{\oplus}} \underset{\underset{NH_2}{|}}{R-CH-CO_2H}$$

Topic Test 1: Structure, Classification, Nomenclature, and Synthesis of Amino Acids

True/False

1. Most naturally occurring amino acids (except Gly, which has H for the side chain) contain a stereogenic center.

2. The amino acid lysine has a basic side chain.

Multiple Choice

3. Which of the following amino acids has a hydrophilic side chain?
   a. Ser
   b. Ala
   c. Phe
   d. All of the above
   e. None of the above

4. Which amino acid has an aromatic side chain?
   a. Trp
   b. Tyr
   c. Phe
   d. All of the above
   e. None of the above

Short Answer

5. Show how one could prepare phenylalanine from a carboxylic acid using the HVZ reaction followed by the Gabriel amine synthesis.

6. The side chain on the amino acid valine is an isopropyl group. Show how one could prepare valine by the Strecker synthesis.

Topic Test 1: Answers

1. **True.** Glycine has the structure H_2NCH_2COOH that does not contain a stereogenic center. All the other naturally occurring amino acids have four different ligands around the alpha carbon and therefore contain stereogenic centers.

2. **True.** Lysine has the side chain $-CH_2CH_2CH_2CH_2NH_2$, which contains the basic amine functional group.

3. **a.** Ser. The hydroxyl of the serine side chain is polar and acts as a hydrogen bond donor or acceptor making it attracted to water (hydrophilic).

4. **d.** All of the above

5.
$$PhCH_2CH_2CO_2H \xrightarrow[\text{2) } H_2O]{\text{1) } Br_2, PBr_3} PhCH_2\underset{Br}{CH}CO_2H \xrightarrow[]{} \xrightarrow[\text{2) } H_3O^{\oplus}]{\text{1) } HO^{\ominus}, H_2O} PhCH_2\underset{NH_2}{CH}CO_2H$$

6.
$$\underset{CH_3}{\overset{CH_3}{>}}CH-\overset{\overset{O}{\|}}{C}-H \xrightarrow[\text{KCN, } H_2O]{NH_4Cl} H_2N-\underset{\underset{CH_3\ \ CH_3}{CH}}{CH}-C\equiv N \xrightarrow[]{H_3O^{\oplus}} H_2N-\underset{\underset{CH_3\ \ CH_3}{CH}}{CH}-CO_2H$$

TOPIC 2: ACID-BASE PROPERTIES OF AMINO ACIDS

KEY POINTS

✓ *What is the structure of a given amino acid at any specified pH?*

✓ *What is a zwitterion?*

✓ *What is an isoelectric point, pI?*

✓ *What are the technique, terminology, and basis for electrophoresis?*

Recall from Chapter 6 that in the presence of base, a carboxylic acid actually exists as a carboxylate ion. We also saw in Chapter 9 that amines are protonated to form alkylammonium ions in the presence of acid. The general structure of an amino acid shown in Table 11.1 is used for convenience but is not entirely accurate because the acid and the base would react to form a dipolar form of the amino acid called a **zwitterion**. The two functional groups of the amino acid, as well as any acidic or basic functional groups on the side chain, will be protonated or deprotonated depending on the pH of the surrounding medium. In acid, everything that can be protonated will be, and in base everything that can be deprotonated will be. This is illustrated below for a simple amino acid assuming that the side chain R is not an acid or base. The pH at which the amino acid has no net charge (i.e., is electrically neutral and exists primarily as the zwitterion) is called the isoelectric point and is abbreviated **pI**.

$$\overset{\oplus}{H_3N}-CH-CO_2H \rightleftharpoons \overset{\oplus}{H_3N}-CH-\overset{\ominus}{CO_2} \rightleftharpoons H_2N-CH-\overset{\ominus}{CO_2}$$
$$\qquad\quad R \qquad\qquad\qquad\quad R \qquad\qquad\qquad\quad R$$

Low pH   Zwitterion   High pH
Acidic   pH near pI   Alkaline

Isoelectronic points for the common amino acids are included in Table 11.1. Differing pI values among amino acids allow them to be separated and identified in a process called **electrophoresis** wherein a mixture of amino acids is buffed to a particular pH and then placed on paper or gel between two electrodes. The charges on the various ionic forms of the amino acids at that pH will cause them to migrate toward the electrodes. Cations migrate toward the negative electrode, whereas anions migrate toward the positive electrode. Zwitterions do not migrate at all because they bear no net charge.

Topic Test 2: Acid-Base Properties of Amino Acids

True/False

1. In an aqueous medium there is no pH at which glycine will exist in the form H₂NCH₂COOH.

2. Most amino acids are conveniently synthesized in the lab by electrophoresis.

Multiple Choice

3. During electrophoresis at pH 3, most amino acids will
   a. be zwitterions.
   b. migrate toward the positive electrode.
   c. have a net positive charge.
   d. All of the above
   e. None of the above

4. The isoelectronic point of an amino acid is the
   a. pH at which all the functional groups are protonated.
   b. same as the pK_a for that acid.
   c. pH at which no net migration occurs during electrophoresis.
   d. All of the above
   e. None of the above

Short Answer

5. Provide a structural formula for serine at its isoelectronic point.

6. Provide a structural formula for the major species present when lysine is at pH 2.

Topic Test 2: Answers

1. **True.** In aqueous systems glycine will be cationic at low pH ($^+H_3NCH_2COOH$), zwitterionic at pH near its isoelectronic point ($^+H_3NCH_2COO^-$), and anionic at high pH ($H_2NCH_2COO^-$).

2. **False.** Electrophoresis can be used to separate and/or identify amino acids but is not a technique for preparing them.

3. **c.** have a net positive charge. In acid the amine is protonated as are the carboxylate and any other basic functional groups. Note also the pI of all amino acids except Asp are above 3.

4. **c.** pH at which no net migration occurs during electrophoresis.

5. $\overset{\oplus}{H_3N}CHCO_2^{\ominus}$
   $\quad\ \ |$
   $\quad\ CH_2OH$

6. $\overset{\oplus}{H_3N}CHCO_2H$
   $\quad\ \ |$
   $\quad CH_2CH_2CH_2CH_2\overset{\oplus}{NH_3}$

TOPIC 3: PEPTIDE BONDS AND SYNTHESIS OF POLYPEPTIDES

KEY POINTS

✓ *What is a peptide?*

✓ *What are di-, tri-, tetra-, oligo-, and polypeptides?*

✓ *What are N-terminal and C-terminal amino acids in a polypeptide?*

✓ *What steps are needed for synthesis of a given oligopeptide from amino acids?*

✓ *What steps are involved in the Merrifield solid phase synthesis?*

The C—N bond of an amide linkage between the amine of one amino acid and the carbonyl of another is called a **peptide bond**. Two, three, four, several, or many amino acids connected in this way are called di-, tri-, tetra-, oligo-, and polypeptides, respectively.

$$\underset{\underset{R^1}{|}}{H_2NCH}\overset{\overset{O}{||}}{C}-\underset{\underset{R^2}{|}}{NHCH}CO_2H \quad = \quad \text{a Dipeptide}$$

Peptide bond

Drawing the complete structures of polypeptides is usually not necessary, and the abbreviations are used to indicate the sequence of amino acids. By convention, the amino acid with the open amine (called the **N-terminal** amino acid) is situated at the left and the open carboxylate (called the **C-terminal** amino acid) is placed on the right as shown in the tripeptide below.

$$\underset{\underset{CH_3}{|}}{H_2NCH}\overset{\overset{O}{||}}{C}-NHCH_2\overset{\overset{O}{||}}{C}-\underset{\underset{CH_2OH}{|}}{NHCH}\overset{\overset{O}{||}}{C}OH \quad = \quad \text{Ala-Gly-Ser} \quad (\underline{not}\ \text{Ser-Gly-Ala})$$

N-Terminal amino acid = Ala; C-Terminal amino acid = Ser

Laboratory synthesis of peptides requires numerous steps to protect (and deprotect) the functional groups that are not intended for reaction. For example, preparing hypothetical dipeptide (AA = amino acid) AA^1-AA^2 from individual amino acids AA^1 and AA^2 would also yield a mixture of AA^2-AA^1, AA^2-AA^2, and AA^1-AA^1 unless steps were taken to ensure that only the amine of AA^2 connected to only the carbonyl of AA^1. Amines are protected by attaching a *tert*-butoxycarbonyl group by reaction with di-*tert*-butyldicarbonate, (BOC)$_2$O. The BOC group can later be removed by trifluoroacetic acid.

$$(CH_3)_3CO-\overset{\displaystyle O}{\overset{\|}{C}}\diagdown_O \quad = \quad (BOC)_2O$$
$$(CH_3)_3CO-\overset{}{C}\diagup^O_{\diagdown O}$$

$$H_2N-\underset{R^1}{CH}-CO_2H \quad \xrightarrow{\qquad} \quad (CH_3)_3CO-\overset{\displaystyle O}{\overset{\|}{C}}-HN-\underset{R^1}{CH}-CO_2H \quad = \quad BOCNH\underset{R^1}{CH}CO_2H$$

$$\text{CF}_3\text{CO}_2\text{H}$$

Carboxylic acids are often protected by Fischer esterification with methyl or benzyl alcohol. The ester can later be converted back into the acid (deprotected) by mild alkaline hydrolysis.

$$H_2N-\underset{R^2}{CH}-\overset{\displaystyle O}{\overset{\|}{C}}-OH \quad \xrightarrow[\text{HCl}]{\text{HOR}} \quad H_2N-\underset{R^2}{CH}-\overset{\displaystyle O}{\overset{\|}{C}}-OR$$
$$(R = CH_3 \text{ or } CH_2Ph)$$

$$\text{H}_2\text{O, NaOH}$$

Peptide bonds between protected amino acids are formed with the help of dicyclohexylcarbodi-imide, DCC. Afterward the protecting groups are removed.

$$BOC-NH\underset{R^1}{CH}\overset{\displaystyle O}{\overset{\|}{C}}-OH \quad + \quad H_2N-\underset{R^2}{CH}-\overset{\displaystyle O}{\overset{\|}{C}}-OR$$

$$\downarrow \text{DCC} \quad = \quad \diagdown\diagup\text{N=C=N}\diagdown\diagup$$

$$BOC-NH\underset{R^1}{CH}\overset{\displaystyle O}{\overset{\|}{C}}-HN-\underset{R^2}{CH}-\overset{\displaystyle O}{\overset{\|}{C}}-OR \quad \left(+ \quad \diagup\diagdown\text{NH}-\overset{\displaystyle O}{\overset{\|}{C}}-\text{NH}\diagup\diagdown \right)$$

$$\downarrow \text{CF}_3\text{CO}_2\text{H}$$

$$\downarrow \text{H}_2\text{O, NaOH}$$

$$NH_2\underset{R^1}{CH}\overset{\displaystyle O}{\overset{\|}{C}}-HN-\underset{R^2}{CH}-CO_2^{\ominus}$$

In summary then, the above synthesis of a dipeptide from two amino acids required five steps:

1. N-Protection of one amino acid with (BOC)$_2$O;
2. C-Protection of another amino acid by esterification;
3. Connect protected amino acids with DCC;
4. Deprotect amine;
5. Deprotect carboxylate.

A larger peptide would require many more steps. A somewhat more practical synthesis of peptides uses similar reactions, but the growing peptide chain is assembled on beads of solid polymeric material. In the Merrifield solid phase synthesis, the intended C-terminal amino acid is first N-protected as usual then attached to the surface of polystyrene that contains about 1% *p*-chloromethyl groups. The solid polymer beads can easily be filtered from reaction mixtures, washed, and put into the next reaction vessel where the BOC group is removed and then another N-protected amino acid is attached using DCC. These steps are repeated until the growing polypeptide is of desired length and is removed from the polymer with hydrogen fluoride, which also deprotects the N-terminal amino acid.

$$\text{BOC-NHCHC(O)-OH} \quad + \quad \text{Cl-CH}_2\text{-[POLY]} \quad \longrightarrow \quad \text{BOC-NHCHC(O)-O-CH}_2\text{-[POLY]}$$
$$R^1 \qquad\qquad\qquad\qquad\qquad\qquad\qquad\qquad\qquad R^1$$

$$\text{BOC-NHCHC(O)-HN-CH-C(O)-O-CH}_2\text{-[POLY]} \xleftarrow[\text{DCC}]{\text{BOC-HN-CH-C(O)-OH, }R^2} \xleftarrow{\text{CF}_3\text{CO}_2\text{H}}$$
$$R^2 \qquad\qquad R^1$$

$\downarrow$ etc. *n*-1 more times $\longrightarrow$

$$\text{BOC-NHCHC(O)}\left(\text{HN-CH-C(O)-O-CH}_2\right)_n\text{-[POLY]}$$
$$R^{n+1} \qquad\qquad R$$

$\downarrow$ HF

$$\text{NH}_2\text{CHC(O)}\left(\text{HN-CH-C(O)-O-H}\right)_n \quad + \quad \text{F-CH}_2\text{-[POLY]}$$
Polypeptide $\quad R^{n+1} \qquad\qquad R$

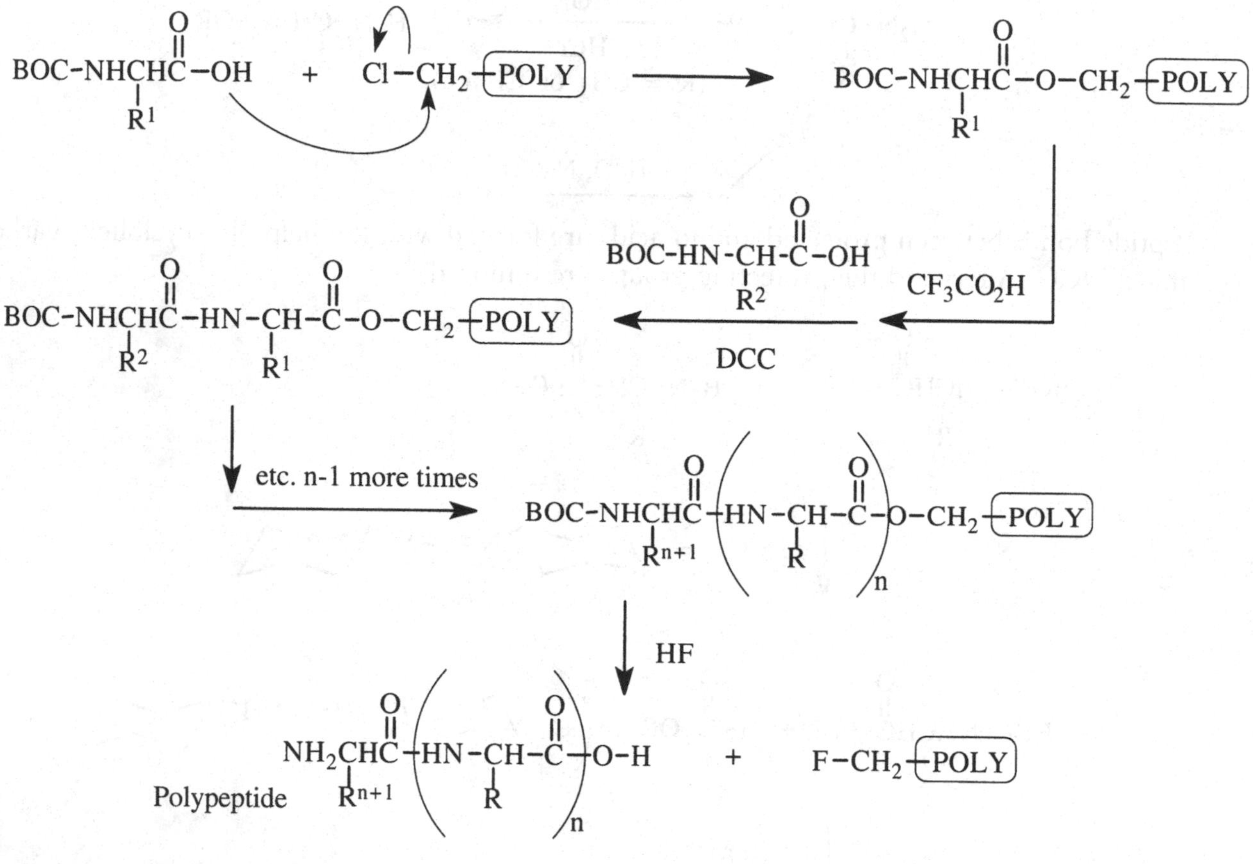

Topic Test 3: Peptide Bonds and Synthesis of Polypeptides

True/False

1. Amine groups are usually protected by trifluoroacetic acid during peptide synthesis.
2. During the Merrifield synthesis, the polypeptide is built up from the N-terminal amino acid to the C-terminal amino acid.

Multiple Choice

3. Reaction of Ala and Phe with DCC would yield which product?
   a. Ala-Phe
   b. Ala-Ala
   c. Phe-Phe
   d. All of the above
   e. None of the above

4. Which of the following conditions will protect the carboxyl group of an amino acid?
   a. CF_3COOH
   b. $(BOC)_2O$
   c. CH_3OH, HCl
   d. DCC
   e. None of the above

Short Answer

5. Draw a complete structural formula for the tripeptide Met-Leu-Pro. Assume the pH is somewhere near neutral.

6. Show the steps one could use to prepare Met-Ala by the Merrifield solid phase technique. Assume that any needed amino acids or protecting agents are available.

Topic Test 3: Answers

1. **False.** Trifluoroacetic acid is often used to remove the BOC protecting group BOC from an amine.

2. **False.** It is the other direction! The polypeptide is built up from the C-terminal amino acid to the N-terminal amino acid during a Merrifield synthesis.

3. **d.** All of the above. Hence the need for protection of some functional groups. Although it is not listed in the problem, Phe-Ala would also likely be produced if no protection were in place.

4. **c.** CH_3OH, HCl. These conditions will convert the carboxylic acid group into a methyl ester by the Fischer esterification. The acid HCl is used dry because the Fischer esterification is reversible and water would promote the reverse reaction, ester hydrolysis.

5.

6.

$$\text{BOC-NHCHC-OH} + \text{ClCH}_2\text{-}\boxed{\text{POLY}} \longrightarrow \text{BOC-NHCHC-O-CH}_2\text{-}\boxed{\text{POLY}}$$

with CH$_3$ groups and carbonyl O shown

↓ CF$_3$CO$_2$H

BOC-NH-CH-C-OH
|
CH$_2$
|
CH$_2$SCH$_3$

$$\text{BOC-CH-C-NH-CH-C-O-CH}_2\text{-}\boxed{\text{POLY}} \longleftarrow[\text{DCC}] \text{NHCHC-O-CH}_2\text{-}\boxed{\text{POLY}}$$

CH$_2$ / CH$_2$SCH$_3$ ... CH$_3$ ... CH$_3$

↓ HF

$$\overset{\oplus}{\text{H}_3\text{N}}\text{-CH-C-NH-CH-C-OH}$$

CH$_2$
|
CH$_2$SCH$_3$ ... CH$_3$

TOPIC 4: DETERMINATION OF POLYPEPTIDE SEQUENCE

KEY POINTS

✓ *What are the steps in the Edman degradation and what information is revealed?*

✓ *Where will trypsin, chymotrypsin, or carboxypeptidase cleave a polypeptide?*

✓ *How can one determine the sequence of amino acids within a polypeptide?*

If a peptide is treated with phenyl isothiocyanate and then mild acid, the N-terminal amino acid is cleaved from the peptide and incorporated into an *N*-phenylthiohydantoin. The reaction scheme is known as the **Edman degradation**.

$$\text{Ph}-\text{N}=\text{C}=\text{S} + \text{H}_2\text{N}-\text{CH}-\overset{\text{O}}{\overset{\|}{\text{C}}}-\text{NH}-\boxed{\text{Polypeptide}} \xrightarrow{\text{H}_3\text{O}^{\oplus}}$$

with R below CH

N-Phenylhydantoin

The R of the N-terminal amino acid is now the R of the product and thus the identity of the N-terminal amino acid is revealed by comparing the properties of the specific *N*-phenylthiohydantoin with those known for derivatives of the common amino acids. Complete sequencing of a polypeptide by repeated Edman degradation is possible in theory but limited by the buildup of unwanted side products. Normally, a large polypeptide will be partially hydrolyzed to smaller fragments that are each sequenced by successive Edman degradations. The entire polypeptide sequence can then be deduced from the fragments. Partial hydrolysis can be carried out with H$_3$O$^+$, leading to essentially random cleavage of peptide bonds. The structure of the polypeptide

Table 11.2 Enzymes for Selective Partial Hydrolysis of Polypeptides	
ENZYME	SELECTIVE PEPTIDE CLEAVAGE AT THE
Trypsin	Carboxyl side of Arg or Lys (basic amino acids)
Chymotrypsin	Carboxyl side of Phe, Tyr, or Trp (aromatic amino acids)
Carboxypeptidase	Amine side of the C-terminal amino acid

can then be deduced from overlapping fragments. Selective partial hydrolysis can be carried out with enzymes that cleave the polypeptide in predictable places. Some of these enzymes are listed in **Table 11.2**.

Topic Test 4: Determination of Polypeptide Sequence

True/False

1. The Edman degradation is a method for selectively cleaving a polypeptide into oligopeptides.

2. The tripeptide Cys-Ser-Gly would be unaffected by the enzyme carboxypeptidase.

Multiple Choice

3. Which tetrapeptide below would yield two dipeptides upon exposure to trypsin?
   a. Met-Lys-Arg-Phe
   b. Arg-Phe-Met-Lys
   c. Met-Arg-Phe-Lys
   d. All of the above
   e. None of the above

4. Which tetrapeptide below would yield two dipeptides upon exposure to chymotrypsin?
   a. Met-Lys-Arg-Phe
   b. Arg-Phe-Met-Lys
   c. Met-Arg-Phe-Lys
   d. All of the above
   e. None of the above

Short Answer

5. A polypeptide was treated with trypsin and yielded the following fragments.
   Asp-Phe-Gly-Ala-Arg
   Cys-Trp-Cys-Lys
   Tyr-Gly-Trp
   Met-Lys
   What is the C-terminal amino acid of this polypeptide?

6. A fresh sample of the same polypeptide from problem 5 was treated with chymotrypsin yielding the following fragments.

Gly-Ala-Arg-Cys-Trp
Met-Lys-Asp-Phe
Cys-Lys-Tyr
Gly-Trp

Deduce the amino acid sequence of the entire polypeptide.

Topic Test 4: Answers

1. **False.** The Edman degradation cleaves off the N-terminal amino acid and incorporates it into an *N*-phenylthiohydantoin.

2. **False.** Carboxypeptidase selectively cleaves off the C-terminal amino acid, which in this case would be Gly.

3. **c.** Met-Arg-Phe-Lys. This tetrapeptide is cleaved to Met-Arg and Phe-Lys upon treatment with trypsin.

4. **b.** Arg-Phe-Met-Lys. This tetrapeptide is cleaved to Arg-Phe and Met-Lys upon treatment with chymotrypsin.

5. **Trp.** All oligopeptide fragments (except the one containing the original C-terminal amino acid) resulting from cleavage with trypsin will necessarily have Lys or Arg as their new C-terminal amino acid. Any observed fragment that has some other amino acid at the C-terminus must have been the C-terminal amino acid from the original polypeptide.

6. Met-Lys-Asp-Phe-Gly-Ala-Arg-Cys-Trp-Cys-Lys-Tyr-Gly-Trp. This sequence can be deduced by matching up the various overlapping fragments given in problems 5 and 6. For example, we know from problem 5 that the C-terminal amino acid is Trp and the last three amino acids at that end are Tyr-Gly-Trp. Because there is only one Tyr in the polypeptide, we conclude by finding it in with the fourth fragment of problem 6 that the last 5 amino acids are Cys-Lys-Tyr-Gly-Trp, and so on.

TOPIC 5: PROTEIN CLASSIFICATION AND STRUCTURE

KEY POINTS

✓ *What are proteins and how are they classified?*

✓ *What are the four levels of protein structure?*

✓ *What forces are responsible for maintaining each level of protein structure?*

✓ *What is the difference between digestion and denaturation of a protein?*

Polypeptides larger than about 50 amino acids are usually called **proteins.** These can be classified according to function, content, or structure. A **simple protein** yields only amino acids upon hydrolysis and a **conjugated protein** contains some nonpolypeptide component. **Globular proteins** are fairly compact and approximately spherical. They are generally water soluble. **Fibrous proteins** are polypeptide strands arranged approximately parallel into bundles that are generally not water soluble. Protein structures are often discussed in terms of levels. The levels of protein structure are summarized below.

Primary structure (1°) is the amino acid sequence along the protein backbone (i.e., the specific order of amino acids). Primary structure is held together by covalent peptide linkages between amino acids. Loss of primary structure requires breaking covalent bonds.

Secondary structure (2°) describes conformations of segments within the protein backbone. Common examples are the **α-helix** and the **β-pleated sheet**. These are held in place by hydrogen bonds between the N—H of the secondary amide and the carbonyl of an amino acid elsewhere in the protein.

Tertiary structure (3°) refers to the further bending or folding of a protein (already in its 2° structure). An analogy is the tangling of a helical telephone cord or "slinky" toy. The helical coil (2°) is still in place but there is now another level of overall shape (3°) superimposed. Tertiary structure is held in place by a variety of forces such as the solubility of the amino acid side chains, salt bridges, and disulfide bridges. When the protein folds up in water, the hydrophilic side chains are toward the outside exposed to water, whereas the hydrophobic side chains are sheltered within. Salt bridges form between charged side groups on amino acids such as Lys (side chain bears —NH_3^+) and Asp (side chain bears —CH_2COO^-) that might be remote on the protein backbone but are drawn together by ionic attraction.

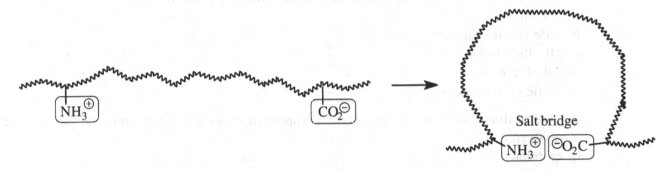

Recall from Chapter 3 that oxidation of thiols yields disulfides. If a disulfide is formed between side chains of cysteine units, those two amino acids will necessarily be close in the protein's native 3° structure even though they may be many amino acids apart in the 1° structure.

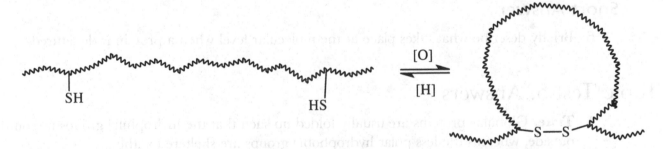

Quaternary structure (4°) is an association or agglomeration of two or more protein subunits (in their 3° structure) held together by noncovalent forces mentioned above including hydrogen bonds, side chain solubility, and salt bridges. A frequently cited example of 4° structure is hemoglobin that is made of four subunits.

Loss of any of the higher order levels of structure (2°, 3°, or 4°) is called **denaturation**. Denatured proteins retain their 1° structure but have unfolded, uncoiled, or separated subunits. Protein **digestion** is the hydrolysis of peptide linkages resulting in loss of 1° structure (and obviously loss of any higher orders of structure as well).

Topic Test 5: Protein Classification and Structure

True/False

1. Globular proteins are generally more water soluble than fibrous proteins.

2. The primary structure of a protein is held in place by hydrogen bonding between adjacent amino acids.

Multiple Choice

3. The α-helix is an example of what level of protein structure?
   a. Primary
   b. Secondary
   c. Tertiary
   d. Quaternary
   e. All of the above

4. Which of the following forces helps to maintain tertiary protein structure?
   a. Salt bridges
   b. Side chain solubility
   c. Disulfide links
   d. All of the above
   e. None of the above

5. A protein that hydrolyzes to yield some component other than only amino acids is called
   a. quaternary.
   b. globular.
   c. simple.
   d. conjugated.
   e. None of the above

Short Answer

6. Briefly describe what takes place at the molecular level when a protein is denatured.

Topic Test 5: Answers

1. **True.** Globular proteins are usually folded up such that the hydrophilic groups are on the outside, whereas the less polar hydrophobic groups are sheltered within.

2. **False.** Primary structure is held in place by covalent peptide (amide) linkages.

3. **b.** Secondary

4. **d.** All of the above

5. **d.** Conjugated

6. A denatured protein has lost some higher level of structure. If subunits separate in a quaternary structure, or a protein backbone unfolds or uncurls from its tertiary or secondary structure, the protein is denatured.

When hair is "permed," the texture (curly or straight) is altered through the breaking and subsequent reforming of disulfide links between cysteine units within the hair's protein stands. The existing disulfide links are first reduced to thiols. Without the extensive cross-linking of disulfide bridges within and among protein strands, the hair is then mechanically manipulated with some appliance such as curlers or rollers to achieve a desired shape. Once the hair is in place, new disulfide bridges are formed that hold the new shape in position.

DEMONSTRATION PROBLEM

A tripeptide was completely hydrolyzed to only yield Ala, Phe, and Val. It did not cleave at all when exposed to chymotrypsin. When treated with phenyl isothiocyanate, the resulting *N*-phenylthiohydantoin contained an isopropyl group. Use these data to deduce the structure of the tripeptide and then show how one could synthesize it from amino acids using the Merrifield solid phase technique.

Solution

There are six possible sequences for a tripeptide containing these three amino acids.

1. Ala-Phe-Val
2. Ala-Val-Phe
3. Phe-Val-Ala
4. Phe-Ala-Val
5. Val-Ala-Phe
6. Val-Phe-Ala

Only structures 2 and 5 would not cleave when treated with chymotrypsin because that enzyme cleaves at the carboxy side of aromatic amino acids and the Phe is the C-terminal amino acid in 2 and 5. The reaction with phenyl isothiocyanante (Edman degradation) shows that Val is the N-terminal amino acid (the isopropyl group observed in the product is the valine side chain). The sequence for the tripeptide must be that shown in 5.

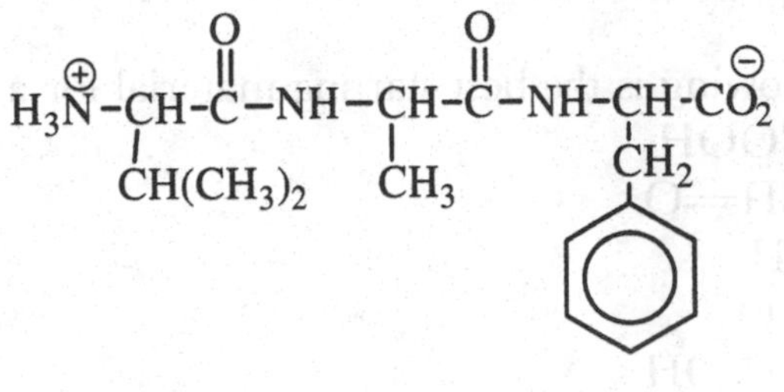

A Merrifield preparation of this tripeptide could be carried out as shown below using N-protected amino acids that are available from the amino acids reacting with (BOC)$_2$O.

$$H_2NCHCO_2H \xrightarrow{\text{(BOC)}_2O} BOC\text{-}NCHCO_2H \quad (R = -CH(CH_3)_2, \ -CH_3, \ -CH_2Ph)$$
$$\overset{|}{R} \qquad\qquad\qquad \overset{|}{R}$$

$$ClCH_2\text{-}\boxed{POLY} \xrightarrow{BOC\text{-}Phe} BOC\text{-}Phe\text{-}O\text{-}CH_2\text{-}\boxed{POLY} \xrightarrow{CF_3CO_2H} \underset{\substack{BOC\text{-}Ala \\ DCC}}{} $$

$$F\text{-}CH_2\text{-}\boxed{POLY} + Val\text{-}Ala\text{-}Phe \xleftarrow{HF} \xleftarrow{BOC\text{-}Val} Ala\text{-}Phe\text{-}O\text{-}CH_2\text{-}\boxed{POLY}$$

Chapter Test

1. Methionine is the amino acid usually involved in forming disulfide bridges.

2. Electrophoresis is a method for attaching the N-terminal amino acid to a growing polypeptide.

3. Fibrous proteins are not generally water soluble.

4. Edman degradation reveals the identity of a peptide's N-terminal amino acid.

Multiple Choice

5. How many fragments will result from cleavage of the peptide Ala-Ser-Arg-Trp-Lys with the enzyme trypsin?
   a. 1
   b. 2
   c. 3
   d. 4
   e. 5

6. Which reagent below is used to remove a polypeptide from the polymer solid support used in the Merrifield synthesis?
   a. CF$_3$COOH
   b. DCC
   c. HF
   d. All of the above
   e. None of the above

7. Which of the following is the best starting material for a Strecker synthesis of serine?
   a. HOCH$_2$CH$_2$COOH
   b. HOCH$_2$CH$_2$CH$=$O
   c. HOCH$_2$COOH
   d. HOCH$_2$CH$=$O
   e. HOCH$_2$CH$_2$CH$_2$OH

8. Which amino acid has a hydrophilic side chain?
   a. Glutamine, Gln

b. Isoleucine, Ile

c. Valine, Val

d. All of the above

e. None of the above

9. Which reaction scheme will lead to alanine?

  a. CH_3CH_2COOH treated with excess ammonia, then DCC, then CF_3COOH.

  b. CH_3CH_2COOH treated with PBr_3 and Br_2, then water, then excess ammonia.

  c. CH_3COOH treated with PBr_3 and Br_2, then water, then excess ammonia.

  d. All of the above

  e. None of the above

10. Which amino acid below would tend to be near the inside of a protein that was folded into its tertiary structure in aqueous solution?

  a. Ser

  b. Asp

  c. Thr

  d. Leu

  e. All of the above

Specify 1°, 2°, 3°, or 4° for the level of protein structure most closely associated with each.

11. Is not lost when a protein is denatured

12. The association or aggregation of two or more polypeptides

13. Held in place exclusively by hydrogen bonds

14. Held together by covalent peptide bonds

15. Maintained by side chain solubility, disulfide links, and salt bridges

16. α-Helix and β-pleated sheet

17. A specific sequence of amino acids

18. Draw the structure of Asp at pH 9

19. Draw tyrosine (Tyr) at its isoelectronic point.

20. Draw the structure of the dipeptide Pro-Gly at pH 3.

Chapter Test: Answers

1. **False**

2. **False**

3. **True**

4. **True**

5. **b** 6. **c** 7. **d** 8. **a** 9. **b** 10. **d**

11. 1°

12. 4°

13. 2°

14. 1°

15. 3°

16. 2°

17. 1°

18. $H_2N{-}CH{-}CO_2^{\ominus}$
    $\quad\quad\;\; CH_2CO_2^{\ominus}$

19. $H_3\overset{\oplus}{N}{-}CH{-}CO_2^{\ominus}$ with CH_2 group attached to a benzene ring bearing OH

20. structure with pyrrolidine ring, $H_2\overset{\oplus}{N}$, $C{=}O$, ${-}NH{-}CH_2\overset{O}{C}OH$

Check Your Performance

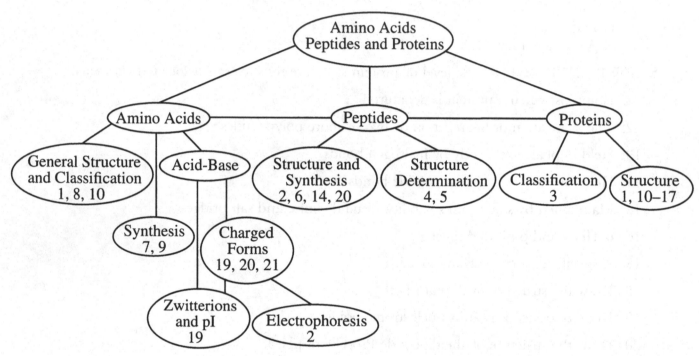

Note the number of questions in each grouping that you got wrong on the chapter test. Identify areas where you need further review and go back to relevant parts of this chapter.

Final Exam

True/False

1. *N*-Methylformamide (*N*-methylmethanamide) will probably show N—H stretch in its infrared spectrum.

2. Aniline has a longer λ_{max} at pH 7 than it does at pH 2.

3. The λ_{max} of 1,3-cyclohexadiene is smaller (shorter wavelength) than that of 1,4-cyclohexadiene or cyclohexene.

4. Tripeptide Ala-Phe-Ala will be cleaved by trypsin.

5. Tripeptide Ala-Ala-Lys will be cleaved by chymotrypsin.

6. The Edman degradation identifies the C-terminal amino acid.

7. The enol tautomer of cyclohexanone is an anion.

8. Attack of a given nucleophile on acetone will be faster than a comparable attack on ethanal.

Multiple Choice

9. What is the order of acidity (most acidic to least acidic)?

$$CH_3CO_2H \qquad CH_3OCH_2CO_2H \qquad CH_3CH_2OH$$
$$\text{I} \qquad\qquad \text{II} \qquad\qquad\qquad \text{III}$$

   a. I, II, III
   b. III, II, I
   c. III, I, II
   d. II, III, I
   e. II, I, III

10. What is the order of reactivity toward nucleophiles (most reactive to least)?

   a. I, II, III
   b. I, III, II
   c. III, II, I
   d. II, I, II
   e. II, III, I

11. What is the order of basicity (most basic to least)?

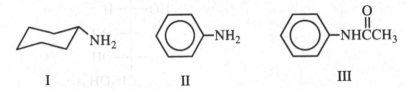

   I   II   III

a. I, II, III
b. I, III, II
c. III, II, I
d. II, I, II
e. II, III, I

Short Answer

Provide unambiguous structural formulas for the following.

12. Isopropyl vinyl ether

13. 4-Mercapto-1-butanol

14. Acetophenone

15. Ethyl β-ketobutyrate (acetoacetic ester)

16. Formaldehyde

17. Phenol

Name the following

18. 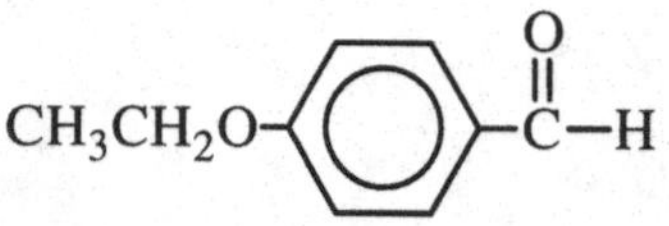 CH_3CH_2O- $\overset{\overset{O}{\|}}{C}-H$

19. $[(CH_3CH_2)_4\overset{\oplus}{N}]\ \overset{\ominus}{NO_3}$

20.

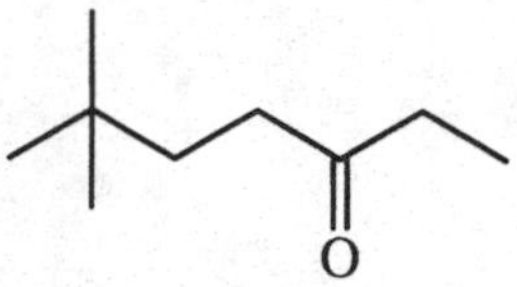

21. 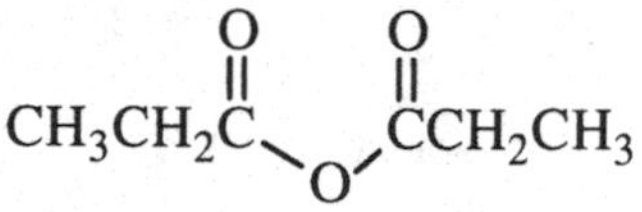 $CH_3CH_2\overset{\overset{O}{\|}}{C}\underset{O}{}\overset{\overset{O}{\|}}{C}CH_2CH_3$

22. $CH_3CH_2CH_2CH_2\overset{\overset{O}{\|}}{C}N(CH_3)_2$

23.

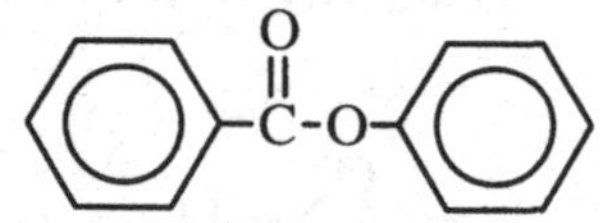

24. Draw all reasonable resonance forms for the anion that results when 4-methyl-1,3-cyclopentanedione is treated with base. Use the correct symbol between resonance forms. (6 points)

25. The structure of D-fructose is shown below. Draw a Haworth projection of D-fructose in its furanose form. Assume the beta anomer.

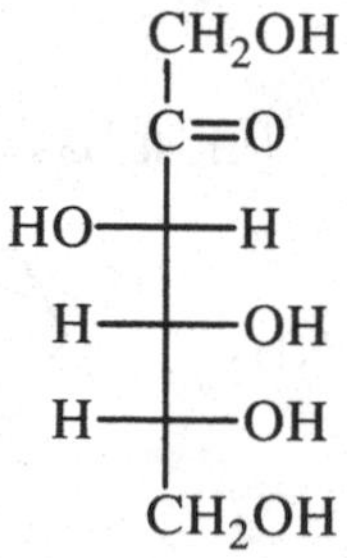

26. Draw a conventional Fischer projection of L-fructose (D-fructose is shown in problem 25). Provide unambiguous structural formulas for the missing organic compounds.

27.
$$\begin{array}{c} CH_3 \\ CHCH_2CH \\ CH_3 \end{array} \overset{O}{\parallel} \quad \xrightarrow[\text{H}_2\text{O}]{\text{NH}_4\text{Cl, KCN}}$$

28.
$$\xrightarrow[\text{mild acid catalyst}]{\text{NH}_2\text{NH}-\bigcirc}$$

29.
$$\xrightarrow{\text{HN}\bigcirc\text{O}}$$

30.
$$\xrightarrow{[(CH_3CH_2)_2Cu]Li} \xrightarrow{H_3O^{\oplus}}$$

31.
$$\begin{array}{c} CH_2 \\ \parallel \\ CH_2CH_2CCH_2CH_2OCH_3 \end{array} \xrightarrow{\bigcirc-\overset{O}{\overset{\parallel}{C}}-O-OH}$$

32.
$$\xrightarrow[-78°\text{ C}]{\substack{\text{1 equivalent} \\ [(CH_3)_2CHCH_2]_2\text{AlH}}} \xrightarrow{H_3O^{\oplus}}$$

33.
$$\xrightarrow[\text{KCN}]{\text{HCN}}$$

34. $CH_3CH_2CH_2\overset{O}{\overset{\parallel}{C}}O-\bigcirc \xrightarrow[\text{NaOH, H}_2\text{O}]{\text{excess}}$

35.
$$\xrightarrow{\text{LiAlH}_4} \xrightarrow{H_3O^{\oplus}}$$

36.
$$\xrightarrow[\text{dry HCl}]{\text{CH}_3\text{OH}}$$

37.
$+ \quad H_2C=C-\overset{O}{\overset{\parallel}{C}}-CH_3 \xrightarrow[\substack{\text{Robinson} \\ \text{annulation}}]{\text{base}}$

38.
$\text{(structure)} \xrightarrow[\text{CrO}_3, \text{H}_2\text{SO}_4, \text{H}_2\text{O}]{\text{excess}}$

39.
$\xrightarrow{\text{HI}}$

40. $\text{HOCH}_2\text{CH}_2\overset{\displaystyle\overset{\text{O}}{\|}}{\text{C}}\text{H} \xrightarrow[\text{CH}_2\text{Cl}_2,\ \text{cold}]{\left[\text{pyridinium}\right]^{\oplus}\ \text{CrO}_3\text{Cl}^{\ominus}}$

41.
$$\underset{\overset{|}{\text{NH}_3^{\oplus}}}{\overset{\overset{\text{CO}_2^{\ominus}}{|}}{\text{CH}}}-\text{CH}_2-\text{S}-\text{S}-\text{CH}_2-\underset{\overset{|}{\text{NH}_3^{\oplus}}}{\overset{\overset{\text{CO}_2^{\ominus}}{|}}{\text{CH}} } \underset{[\text{O}]}{\overset{[\text{H}]}{\rightleftharpoons}}$$
$\left(\begin{array}{c}\text{Similar to reactions}\\\text{of hair texture treatments}\end{array}\right)$

42.
$\xrightarrow[\text{HOCH}_2\text{CH}_3]{\text{Na}^{\oplus}\ ^{\ominus}\text{OCH}_2\text{CH}_3}$

43.
$\xrightarrow{\text{HOCH}_2\text{CH}_3,\ \text{H}^{\oplus}}$

Show the reagents and/or conditions one could use to carry out the following transformations. More than one step may be required.

44.
$\longrightarrow$

45. $\text{BrCH}_2\text{CH}_3 \longrightarrow \left[\text{Ph}_3\overset{\oplus}{\text{P}}-\overset{\ominus}{\text{C}}\text{HCH}_3 \longleftrightarrow \text{Ph}_3\text{P}=\text{CHCH}_3\right]$

46.
$\longrightarrow$

47. $\text{CH}_3\text{CH}_2\text{CO}_2\text{H} \longrightarrow \text{CH}_3\underset{\overset{|}{\text{NH}_2}}{\text{CH}}\text{CO}_2^{\ominus}$

48. $\text{CH}_3\overset{\overset{\text{O}}{\|}}{\text{C}}\text{CH}_2\overset{\overset{\text{O}}{\|}}{\text{C}}\text{OCH}_2\text{CH}_3 \longrightarrow \text{CH}_3\overset{\overset{\text{O}}{\|}}{\text{C}}\text{CH}_2\text{CH}_2\text{CH}_3$

49. A compound with the formula C_4H_8O shows a strong carbonyl stretching peak in the infrared spectrum near $1715\,\text{cm}^{-1}$. It produces no visible result when treated with silver oxide in aqueous ammonia (Tollens' reagent). Propose a structure.

Use the spectral data provided to deduce unambiguous structural formulas for the following compounds.

50. The mass spectrum of this hydrocarbon shows a molecular ion at 120 amu. The ^{13}C-nuclear magnetic resonance (NMR) spectrum (proton noise decoupled) showed only six lines. The ^{1}H-NMR spectrum is summarized below.
    7.25 ppm (broad singlet, 5H)
    2.90 ppm (septet, 1H)
    1.22 ppm (doublet, 6H)

51. Formula = $C_4H_8O_3$. The infrared spectrum shows a strong peak near 1710 cm^{-1} and a strong broad peak at 2500 to 3100 cm^{-1}. The cm^1 H-NMR spectrum is summarized below.
    12.0 ppm (broad singlet, 1H)
    4.15 ppm (singlet, 2H)
    3.60 ppm (quartet, 2H)
    1.20 ppm (triplet, 3H)

Write out reaction schemes for the following reactions.

52. The acid-catalyzed reaction of 5-hydroxy-2-pentanone to yield a cyclic hemiacetal.

53. The acid-catalyzed Fischer esterification between acetic acid and ethanol.

54. The hydroxide-catalyzed aldol reaction of acetone.

55. The dehydration of 2-methylcyclohexanol with concentrated sulfuric acid.

Answers

1. **True**
2. **True**
3. **False**
4. **False**
5. **False**
6. **False**
7. **False**
8. **False**
9. **e**
10. **b**
11. **a**

12.

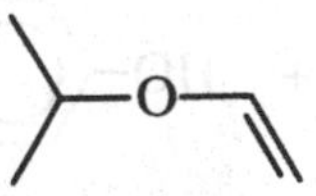

13. HSCH$_2$CH$_2$CH$_2$CH$_2$OH

14.

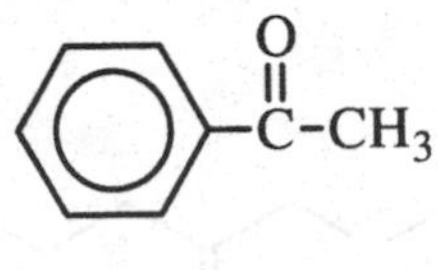

15. CH$_3$CCH$_2$COCH$_2$CH$_3$ (with two C=O)

16.

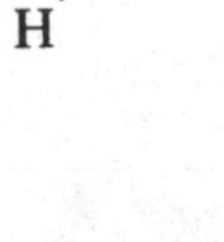

17.

18. *p*-Ethoxybenzaldehyde or 4-ethoxybenzaldehyde

19. Tetraethylammonium nitrate

20. 6,6-Dimethyl-3-heptanone

21. Propanoic anhydride

22. *N,N*-dimethylpentanamide

23. Phenyl benzoate

24.

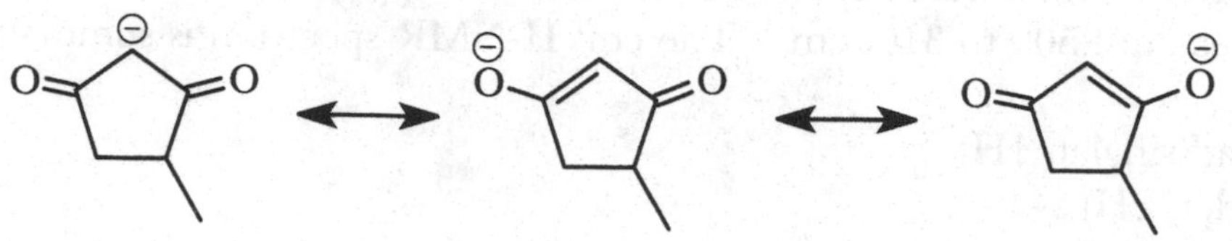

25.

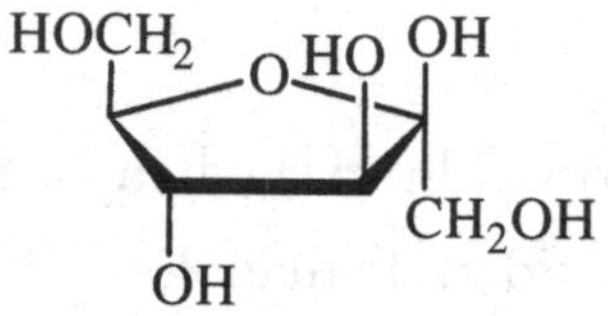

26.

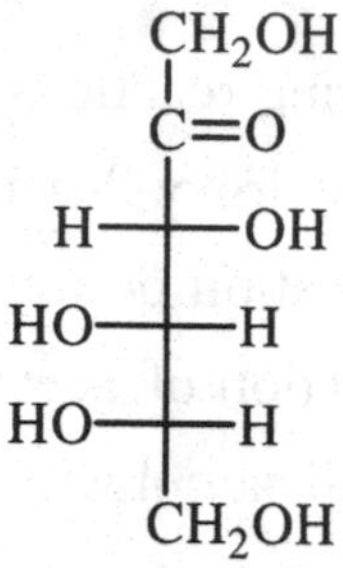

27.

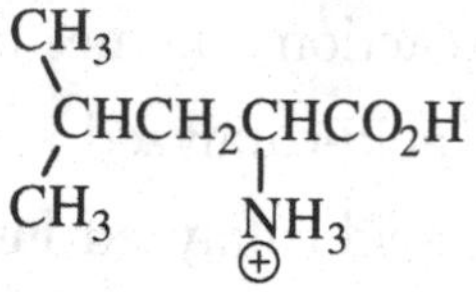

28.

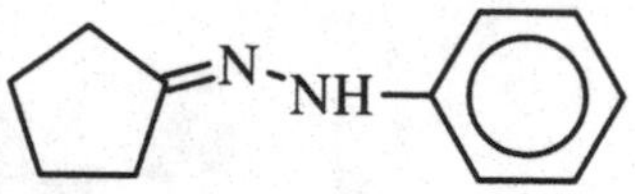

29.

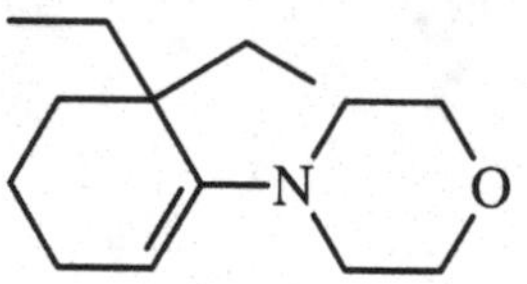

30.

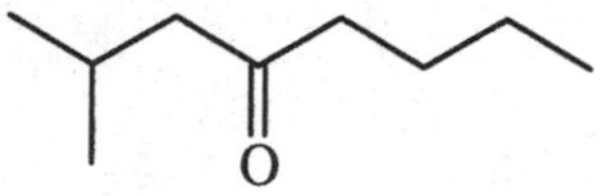

31.

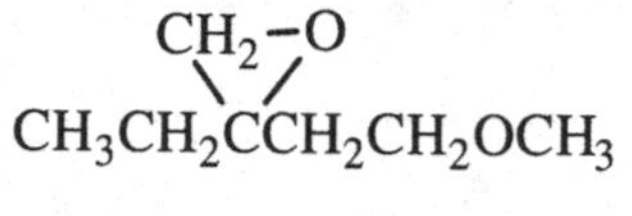

32.

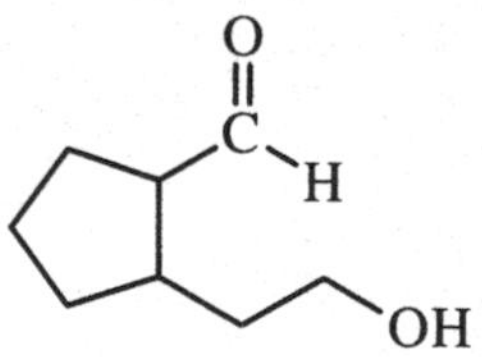

33.

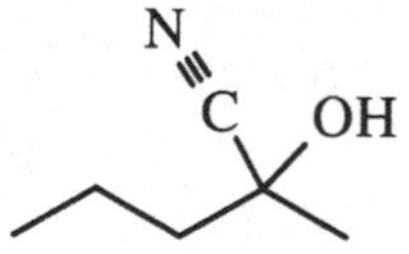

34. 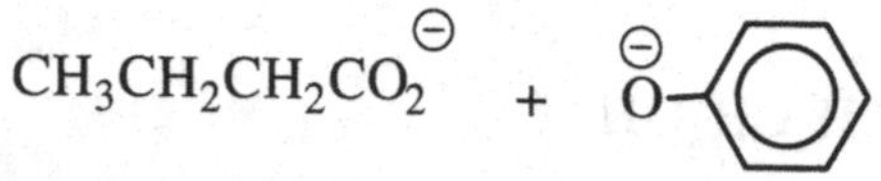 $CH_3CH_2CH_2CO_2^{\ominus}$ +

35. 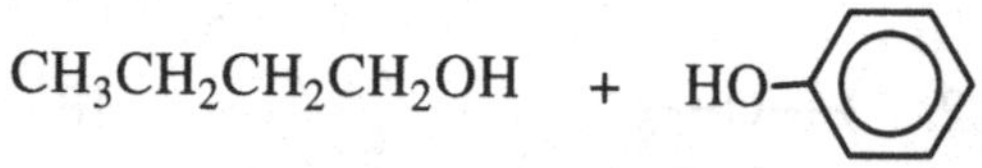 $CH_3CH_2CH_2CH_2OH$ + $HO-$

36.

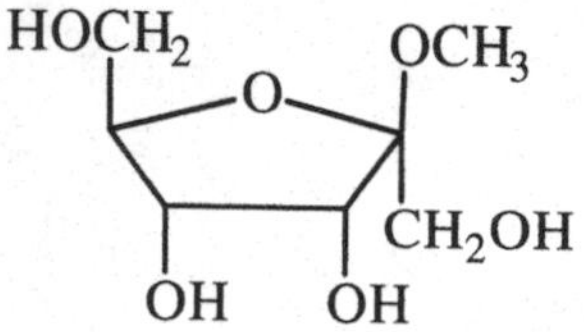

37.

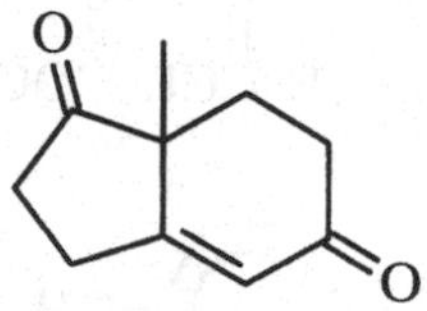

38.

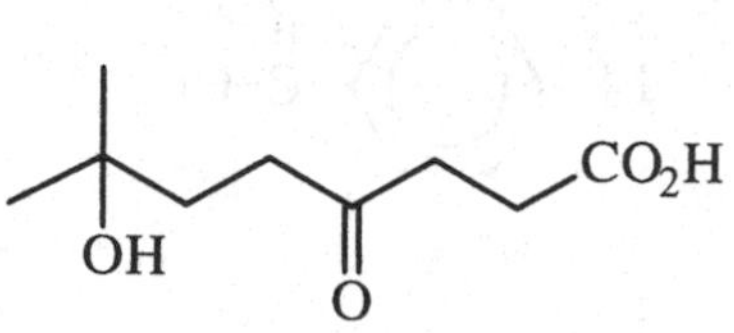

39. 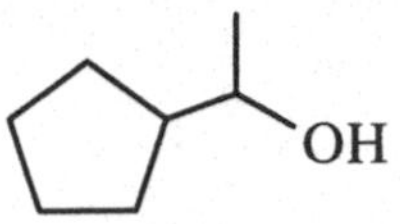 cyclopentyl-CH(CH₃)-OH + $ICH_2CH_2CH_3$

40. $\overset{O}{\overset{\|}{H}C}CH_2CH_2\overset{O}{\overset{\|}{C}}H$

41. $2\overset{\overset{\displaystyle CO_2^{\ominus}}{|}}{\underset{\underset{\displaystyle NH_3^{\oplus}}{|}}{C}}HCH_2SH$

42.

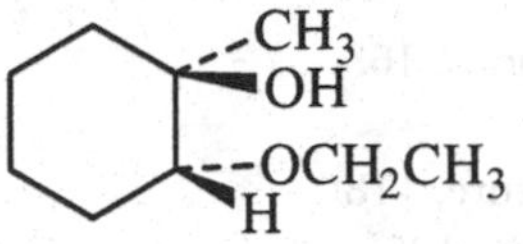

43.

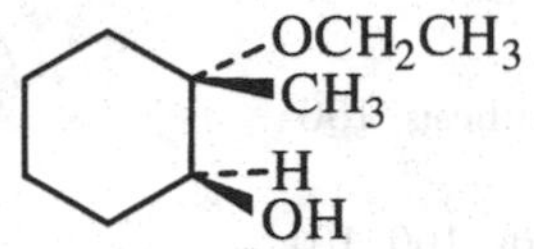

44. Hg(O₂CCF₃)₂, HOCH₃ then NaBH₄ (other methods possible including Markovnikov hydration to 2-pentanol then Williamson ether synthesis with methyl iodide).

45. Ph₃ then BuLi.

46. Ph₃P=CHCH₃ (other answers possible including reaction with ethyl Grignard then dehydrating the symmetrical tertiary alcohol).

47. Br₂, PBr₃ then excess NH₃ (another reasonable answer is HVZ then Gabriel synthesis).

48. CH₃CH₂ONa, then CH₃CH₂Br, then heat with H₃O⁺.

49. $CH_3CH_2\overset{O}{\overset{\|}{C}}CH_3$

50.

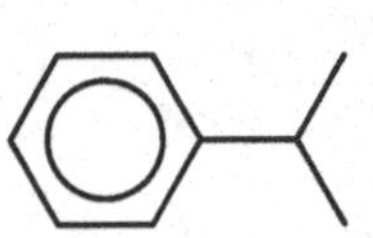

51. $CH_3CH_2OCH_2CO_2H$

52. $HOCH_2CH_2CH_2\overset{O}{\overset{\|}{C}}CH_3 \xrightarrow{H^\oplus}$

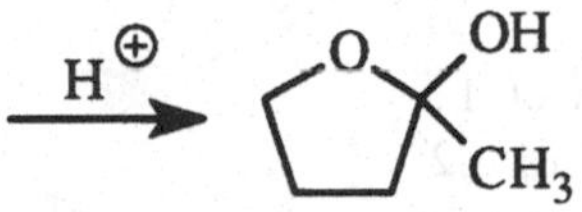

53. $CH_3CO_2H + HOCH_2CH_3 \underset{}{\overset{H^\oplus}{\rightleftharpoons}} CH_3CO_2CH_2CH_3 + H_2O$

54. $2CH_3\overset{O}{\overset{\|}{C}}CH_3 \xrightarrow{OH^\ominus} CH_3\overset{O}{\overset{\|}{C}}CH_2\underset{\underset{\displaystyle CH_3}{|}}{\overset{\overset{\displaystyle OH}{|}}{C}}CH_3$

55.

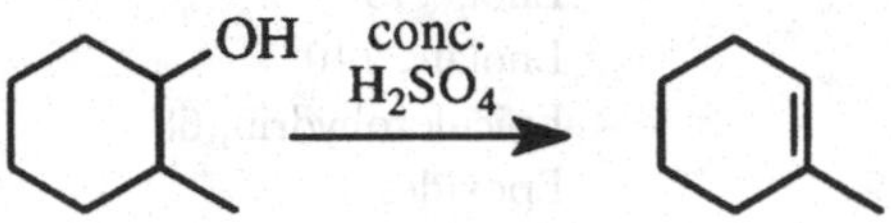

INDEX